Receptors and Recognition

General Editors: P. Cuatrecasas and M.F. Greaves

About the series

Cellular recognition — the process by which cells interact with, and respond to, molecular signals in their environment — plays a crucial role in virtually all important biological functions. These encompass fertilization, infectious interactions, embryonic development, the activity of the nervous system, the regulation of growth and metabolism by hormones and the immune response to foreign antigens. Although our knowledge of these systems has grown rapidly in recent years, it is clear that a full understanding of cellular recognition phenomena will require an integrated and multidisciplinary approach.

This series aims to expedite such an understanding by bringing together accounts by leading researchers of all biochemical, cellular and evolutionary aspects of recognition systems. The series will contain volumes of two types. First, there will be volumes containing about five reviews from different areas of the general subject written at a level suitable for all biologically oriented scientists (Receptors and Recognition, series A). Secondly, there will be more specialized volumes, (Receptors and Recognition, series B), each of which will be devoted to just one particularly important area.

Advisory Editorial Board

Receptors and Recognition

Receptors and
Recognition

Series B Volume 4

Specificity of Embryological Interactions

Edited by
D. R. Garrod

Department of Biology,
University of Southampton, U.K.

LONDON
CHAPMAN AND HALL

A Halsted Press Book
John Wiley & Sons, New York

First published 1978
by Chapman and Hall Ltd.,
11 New Fetter Lane, London EC4P 4EE

© *1978 Chapman and Hall*

Typeset by C. Josée Utteridge-Faivre
and printed in Great Britain
at the University Printing House, Cambridge

ISBN 0 412 14420 4

Distributed in the U.S.A. by Halsted Press
a Division of John Wiley & Sons, Inc., New York

Contents

Contributors

S.H. Barondes, Department of Anatomy, School of Medicine, University of California, San Francisco, California, U.S.A.

A.S.G. Curtis, Department of Cell Biology, University of Glasgow, Glasgow, U.K.

R.M. Gaze, National Institute for Medical Research, Mill Hill, London, U.K.

J. Hermolin, Department of Zoology, University of Wisconsin, Madison, Wisconsin, U.S.A.

J. Lilien, Department of Zoology, University of Wisconsin, Madison, Wisconsin, U.S.A.

P. Lipke, Department of Zoology, University of Wisconsin, Madison, Wisconsin, U.S.A.

D.M. Noden, Department of Zoology, University of Massachusetts, Amherst, Massachusetts, U.S.A.

S.D. Rosen, Department of Anatomy, School of Medicine, University of California, San Francisco, California, U.S.A.

M.S. Steinberg, Department of Biology, Princeton University, Princeton, New Jersey, U.S.A.

R.S. Turner, Jr., Department of Biology, Wesleyan University, Middletown, Connecticut, U.S.A.

Preface

The Specificity of Embryological Interactions is a very broad title which could relate to many aspects of the developmental process. I must begin, therefore by stating that this book is about cellular interactions in development, particularly those cellular interactions which result in the precise positioning of cells within embryos as a result of morphogenetic movements.

Let us take a well-known example, that of embryonic gastrulation, to illustrate the nature of the problem. In amphibian development, a hollow ball of cells, the blastula, arises from the egg by a process of cell division or cleavage. During cleavage, there is no spatial rearrangement of the cytoplasmic material of the egg; it is merely packaged into small units, each unit or cell becoming separated from its neighbours by the formation of its surface membrane. The pigmentation pattern of the blastula and the distribution of cytoplasm is identical to that of the egg from which it was derived. The three germ layers, the ectoderm, mesoderm and endoderm are on the surface of the blastula.

Then, quite suddenly, the situation changes; gastrulation begins and the germ layers are on the move. For the embryo, gastrulation involves a massive and dramatic rearrangement of its cells. Two of the germ layers, the mesoderm and endoderm, move inside while the third, the ectoderm, spreads to cover them. Thus roughly half the surface cells of the blastula move inside during gastrulation. As they move inside the ectoderm, the mesoderm and endoderm generate the three-layered organisation fundamental to all later stages of development with endoderm inside, ectoderm outside and mesoderm in between. The correct positioning of the germ layers during gastrulation is crucial to the normal progress of the development of the organism.

There are numerous other well-described morphogenetic movements as a result of which cells take up particular postions within embryos, for example the migration of neural crest cells, the formation of nerve connections, the passage of primordial germ cells to the gonads and the movements of mesodermal cells to form the heart. The precision and accuracy, both spatial and temporal, with which these movements are accomplished, has lead to the liberal use of the word 'Specific' to describe both the movements and the cellular interactions they involve. In his paper, 'The problem of specificity in growth and development', Paul Weiss defined 'selectivity' (or 'specificity' − he used the words synonymously) as 'the faculty of a process or of a substance to activate, to alter the state of, or to combine with, certain elements in preference to, and to the exclusion of, other elements of the same system'. If we accept this thoroughly reasonable definition, it seems quite justifiable to refer to the observed positioning of cells and tissues as a result of morphogenetic movements as selective or specific. Two sentences later, however, Weiss cautions us. Thirty years later it seems relevant to repeat his statement: 'By custom, however, it [i.e. the word 'specificity'] has acquired a secondary meaning signifying those properties of each system which make selectivity of interaction possible'. Good scientists should not fall prey to such customs. We should recognize that in describing cell positioning in development as 'specific', we are, in the first instance, doing no more than *describe* an aspect of cell behaviour. It is one thing – in my view a justifiable thing – to describe cell behaviour in these terms. It is quite another thing – in my view a dangerous thing – to move incautiously from the description of cell behaviour to making deductions about molecular mechanisms which might be responsible for that behaviour.

Our consideration of the problem begins with two chapters which directly present and consider examples of cell positioning in development. Some reviewers have grouped positioning processes together, thus implying that different examples may involve similar mechanisms. It is worth stressing therefore that different types of morphogenetic movements are well known to involve different types of mechanisms. Thus, our first example, the migration of neural crest cells involves the migration of individual cells or small groups of cells to a variety of different places in the embryo. The majority of morphogenetic movements, though different in detail from neural crest migration, are of the same general type, involving the repositioning by migration of individual cells, groups of cells or whole tissues. Our second example, the formation of specific nerve connections, is a somewhat different type of process. Instead of moving from one place to another, the nerve cell bodies remain stationary but grow axons, the tips of which migrate to specific sites as they grow, resulting in the formation of extremely intricate and precise connection patterns. At least one other type of general mechanism comes to mind, exemplified by the positioning of primordial germ cells of birds. These originate outside the embryo proper and pass into the embryonic circulation. They then settle on the endothelial lining of blood vessels adjacent to the prospective gonads before migrating to the germinal ridges. Cell positioning by such a mechanism is probably rare in development, but would also seem to be involved in the settling of lymphocytes in adults and possibly in the metastasis of tumour cells.

The following five chapters require only the briefest of justifications. They may be divided into two groups. Chapters 3, 4 and 5 may be loosely entitled '*In vitro* studies on the selective adhesion of embryonic cells'. Three rather different approaches to the understanding of the mechanistic basis of selective cell adhesion are presented so that the validity and achievements of each may be stressed. The last two chapters deal with the use of 'model systems', i.e. sponges and cellular slime moulds, to study molecular mechanisms of cell adhesion. Some impressive advances have been made in these areas in the last few years, demonstrating the suitability of these simple organisms for basic studies on complex developmental phenomena.

In summary, a broad approach to the problem of selective adhesion and cell positioning in development has been taken. A variety of experimental approaches are considered while the descriptive basis of cell positioning processes in development are kept in mind.

February, 1978 *D.R. Garrod*

Part 1: Specific Cell Positioning in Embryonic Development

1 Interactions Directing the Migration and Cytodifferentiation of Avian Neural Crest Cells

DREW M. NODEN

Acknowledgements

I wish to thank K. Tosney for generously providing the scanning electron micrographs used in this paper. Original research was supported by BSSG Grant 07048 to the University of Massachusetts.

Specificity of Embryological Interactions
(*Receptors and Recognition,* Series B, Volume 4)
Edited by D.R. Garrod
Published in 1978 by Chapman and Hall, 11 New Fetter Lane, London EC4P 4EE

1.1 INTRODUCTION

Interactions between developing cells and their microenvironment, including other cells, have been recognized for nearly a century as essential correlates of morphogenesis and cytodifferentiation in multicellular organisms. Originally, these interactions were thought of as a series of discrete embryonic inductions in which a pluripotential, competent tissue became committed to a specific fate (determined) in response to an instructional stimulus emanating from an inducing tissue (reviewed by Spemann, 1938).

Currently, we believe that the embryonic determination and subsequent development of most cells is the result of a *continuous* interplay between each cells and its microenvironment. Thus, rather than focussing on a single interaction, it is necessary to catalogue and characterize all of the components, both cellular and extracellular, of the continually changing milieu around each developing cell, and then assess the role of these components, acting alone or in concert, in affecting the development of the cell.

The vertebrate neural crest has recently become a popular tissue for analysis of these problems. Arising from the neural folds as they fuse to form the brain and spinal cord, neural crest cells migrate extensively into many regions of the embryo. These morphogenetic movements are highly patterned and expose crest cells to many cellular and extracellular constituents within the embryo. Subsequently, neural crest cells contribute to a wide variety of diverse tissues, including sensory and autonomic ganglia, cranial cartilages and bones, the cornea, pigment, and numerous others (see Table 1.1).

The embryonic origin of this population is well defined both spatially and temporally, which has enabled many investigators to study the development of crest cells using the classical microsurgical methods of experimental embryology (reviewed by Hörstadius, 1950; Weston, 1970; Andrew, 1971). These studies have provided detailed descriptive accounts of the migration and cytodifferentiation of most amphibian and avian neural crest cells, and revealed the essential role played by interactions between these cells and their environment in neural crest cell development. With this background it has recently been possible to begin characterizing these environmental components and analyzing their influence in specifying the migratory patterns and cytodifferentiation of neural crest cells.

The following discussion is based largely on experimental analyses using avian neural crest cells. However, this reflects the author's bias and is not meant to diminish the importance of the many elegant and incisive experiments performed on amphibian embryos, most of which have been reviewed by Hörstadius (1950) and Weston (1970).

5

Table 1.1 Major derivatives of the avian neural crest

---

1. *Neural derivatives*

 Sensory ganglia (including neurons, glia, sheath, and satellite cells)
  trigeminal[P] (V)
  root[P] (VII)
  superior (IX)
  jugulare (X)
  spinal dorsal root ganglia

 Sympathetic ganglia (non-vascular components only) and tissues
  paravertebral (chain)
  prevertebral (coeliac, mesenteric, adrenal, retro-aortic complexes)
  adrenal medulla

 Parasympathetic ganglia (non-vascular components only)
  ciliary
  submandibular, ethomid, sphenopalatine, otic, lingual
  intrinsic visceral
  Meissner's, Auerbach's, Remak's, pelvic plexes

 Schwann sheath cells

 Suporting cells (glia, satellite cells) but not neurons
  geniculate (VII) and acoustic (VIII) ganglia[PP]
  petrosal ganglion[PP] (XI)
  nodose ganglion[PP] (X)

 [P], [PP] some ([P]) or all ([PP]) of these neurons are of epidermal placode origin

2. *Cartilages, bones, muscles, and connective tissues (from cranial neural crest only)*

 Visceral arch carilages*
  1st arch: Meckel's quadrate
  2nd arch: columella, stylohyal, basihyal (entoglossal)
  branchial arches: all remnants of basi-, cerato- and epi-branchials

 Loose connective tissue of face, tongue, and lower jaw

 * structures listed are those found in the embryo either prior to replacement
 by bone (*) or before fusion of ossifield skeletal elements (†)

3. *Other neural crest derivatives*

 Pigment cells
  melanocytes of dermis, mesenteries, internal organs, epidermis, etc.
  melanophores of iris

 Secretory cells
  carotid body Type I cells
  calcitonin-producing cells of ultimobranchia body

Table 1.1 Major derivatives of the avian neural crest (*continued*)

---

3. *Other neural crest derivatives* (*cont.*)

Chondrocranial cartilages*
ethmoid, interorbital septum
anterior and posterior orbitals
sclera, nasal capsule

Lower jaw bones†
dentary, angular, supra-angular, opercular (splenial)
articular, quadratojugal

Upper jaw, palatal, and cranial vault bones†
maxilla, premaxilla, palatine, nasal, prefrontal, lacrimal, frontal
(rostral part)
anterior parasphenoid (rostrum), pterygoid, squamosum (squamous
temporal)
scleral ossicles

Muscles
ciliary muscles (striated)
some cranial vascular and dermal smooth muscles (see below)
visceral arch-derived muscles (minor component)

Corneal endothelium and stromal fibroblasts

Mesenchymal component of adenohypophysis, lingual gland, parathyroid,
thymus, thyroid

Dermis and sub-cutaneous adipose of face, jaw, and upper neck, and
mesenchyme adjacent to oral epithelium

Leptomeninx of the diencephalon and telencephalon

Arterial wall smooth muscle and elastic fiber tissues, but NOT endothelial
layer derivatives of visceral arch vessels, including parts of internal,
external, and common carotids, 4th arch component of systematic
aorta, pulmonary (6th arch) vessels

Venous wall components, exlusing endothelium in superficial facial, oral and
jaw regions

1.2 EARLY MIGRATORY BEHAVIOR AND PATTERNS

1.2.1 Formation and onset of migration

Neural crest cells first appear in the chick embryo as a wedge of cells between the
superficial epidermis and the neuro-epithelium of the apposed mesencephalic neural
folds (Fig. 1.1a). The time of their initial formation throughout pre-otic regions of

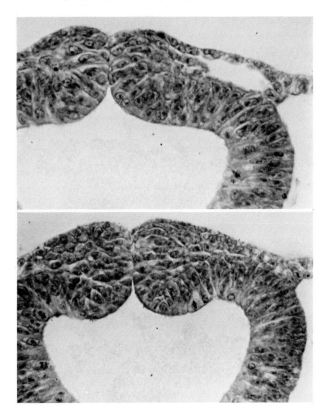

Fig. 1.1a,b Two stages in the initial closure of mesencephalic neural folds
and formation of the neural crest. Note the close apposition of superficial
ectoderm to the dorso-lateral mesencephalic neuro-epithelium.

the head is correlated similarly with neural fold apposition (Holmdahl, 1928;
Hillman and Hillman, 1965; di Virgilio *et al.,* 1967). In the trunk and post-otic head
regions, crest cells do not form for several hours after neural tube closure (Weston,
1963; Bancroft and Bellairs, 1976), appearing in synchrony with, or slightly in
advance of, adjacent somite formation (Tosney, 1978).

Scanning and transmission electron microscopy of the formation of trunk neural
crest cells reveals that most of the cells are initially elongated in the dorso-ventral
plane, perpendicular to the roof of the spinal cord (Bancroft and Bellairs, 1976;
Tosney, 1978). Head crest cells appear to be similarly elongated perpendicular to the
germinal neuro-epithelium, although these would lie in a medio-lateral plane due to
their relatively precocious appearance (Fig. 1.1b). As the crest population increases,
both by additional contribution from the neuro-epithelium and intrinsic proliferation,
the amount of extracellular space diminishes and the cells become flattened between

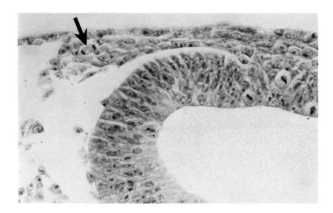

Fig. 1.2 Early migration of mesencephalic neural crest cells. The leading cells are not in contact with either the superficial ectoderm or the neuro-epithelium. The arrow indicates a dividing cell.

the roof of the neural tube and the overlying epidermis. At this stage most of the cells are not oriented in any particular direction, although some at the lateral margins may be elongated in the medio-lateral plane.

As migration begins the cell density decreases in the lateral aspect of the neural crest population. Cells at the leading edge are polarized with their long axes in the direction of migration (Fig. 1.3). For the most part these leading cells are not in contact with the basal lamina of either the neuro-epithelium or the epidermis (Ebendal, 1977; Tosney, 1978), as shown in Fig. 1.2. While transplantation studies suggest that a few individual cells emigrate in advance of the major population (Weston, 1963; Noden, 1975), examination of crest cells *in situ* indicates that this number is small (Bancroft and Bellairs, 1976; Ebendal, 1977; Tosney, 1978). Often a cell may break away from the irregular leading edge of the population (Fig. 1.3), but such breaks appear to be temporary at this stage.

1.2.2 Patterns of neural crest cell migration

Appreciation of the diverse yet precisely delimited patterns of avian neural crest cell migration was incomplete until recently due to a lack of reliable morphological criteria by which these cells could be distinguished from mesodermal cells. However, by autoradiographically studying the distribution of ^{3}H-thymidine-labeled chick neural crest cells which had previously been transplanted orthotopically into an unlabeled host embryo, Weston (1963), Johnston (1966) and, later, Noden (1973, 1975) have been able to map precisely the migratory routes of these cells.

Weston's analysis revealed that trunk crest cells migrate towards the somite as an unsegmented mesenchyme, which confirms Detwiler's (1937) observations on early

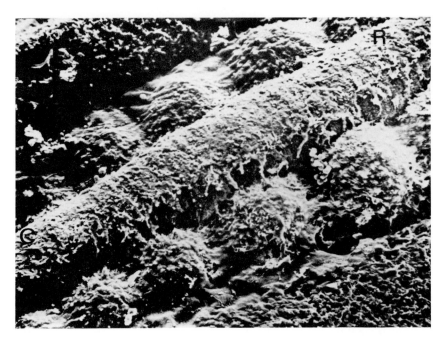

Fig. 1.3 Scanning EM showing the early migration of chick thoracic neural crest cells. The superficial ectoderm has been removed to reveal underlying tissues. R and C indicate rostral and caudal aspects. (Reproduced with permission from K. Tosney (1978) and Academic Press.)

neural crest migration in amphibian embryos. Either before (Weston, 1963) or upon reaching (Tosney, 1978) the dorsal margin of the somite, the crest population becomes split into two parts. A relatively small portion of the population continues moving laterally between the somitic mesoderm and the surface ectoderm (Rawles, 1948; Teillet, 1971). These cells, most of which are pigment cell precursors, form a sparse monolayer (Tosney, 1978) which later becomes interspersed with cells of dermatomal origin. Movement into the epidermis commences on the fourth day of incubation (Le Douarin and Teillet, 1970).

In contrast, most trunk crest cells migrate ventrally as part of a multilayered mesenchyme and enter a region bounded by the neural tube, medially, and the dorsal somitic mesenchyme, laterally. Some of these cells are closely apposed to each other and form small aggregates (Bancroft and Bellairs, 1976; Ebendal, 1977), while others are in physical contact with surrounding components only via long filopodia. As the leading cells move ventrally, contact is established with the sclerotomal region of the somite, which is only partially covered with a basal lamina (Tosney, 1978). Subsequently, the crest and sclerotome populations become interspersed. Migrating cells which initially enter the inter-somitic region later move into

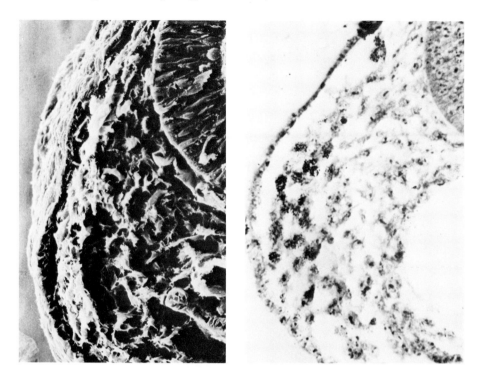

Fig. 1.4a,b The later migration of mesencephalic neural crest cells.
(a) A wedge of crest cells can be seen between the superficial ectoderm and
the mesodermal mesenchyme. (b) shows autoradiographically the location
of [3] H-thymidine-labeled neural crest cells which had been transplanted in
the place of mesencephalic neural fold tissues of an unlabeled chick host
16 hours prior to fixation. (a) Courtesy of K. Tosney; (b) From Noden,
1975).

the adjacent sclerotomal mesenchyme (Weston, 1963), resulting in a metameric
distribution of trunk crest cells.

Shortly thereafter, loose aggregates of neural crest cells appear within the somites
indicating that some members of the population have ceased their ventrally directed
migration and begun to form dorsal root ganglia. Others leave the somitic mesenchyme
and move ventrally towards the dorsal aorta. From this latter population will develop
sympathetic ganglia and, in the posterior thoracic (somite levels 18–24; Le Douarin
and Teillet, 1971) levels, adrenal medullary cells. In addition, some of these deeply
migrating crest cells develop into pigment cells found along mesenteries and the
peritoneum, and also associated with several internal organs, especially the
mesonephori and gonads. Some lumbo-sacral crest cells form parasympathetic
neurons of the large intestine (Le Douarin and Teillet, 1973).

Avian cephalic neural crest cells display migratory patterns which are strikingly

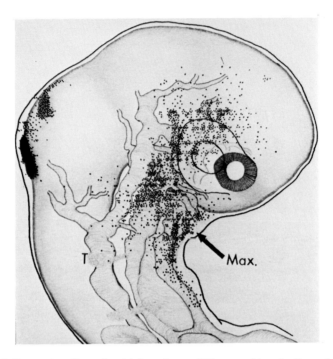

Fig. 1.5 Reconstruction of a chick embryo, 53 hours of incubation whose mesencephalic neural folds were replaced with homologous tissue from a donor chick pre-incubated with ^{3}H-thymidine. Black dots show the location of all labeled cells in the host 22 hours after the transplant. Max. indicates the future maxillary process. (From Noden, 1975).

different from those just described. In addition there exist extensive differences in migratory behavior between mesencephalic, rhombencephalic, and prosencephalic crest cells (Noden, 1973, 1975).

Upon leaving their origin, most mesencephalic crest cells migrate laterally (Fig. 1.2) ventrally (Fig. 1.4) between the presumptive epidermis and the mesodermal mesenchyme. The mitotically-active crest population maintains a multilayered mesenchymal appearance as the leading edge approaches the oropharyngeal region. Transplantation experiments by Johnston (1966) and Noden (1975) clearly indicate that the neural crest cells remain segregated from mesodermal mesenchyme during this early migratory phase. The entire complement of mesencephalic crest cells subsequently moves away from the midbrain and forms the maxillary process, with some contribution to periocular and mandibular mesenchyme. This migratory pattern is shown schematically in Fig. 1.5.

The behavior of crest cells emigrating from the anterior rhombencephalon differs from that seen in the midbrain. As migration commences this population forms a

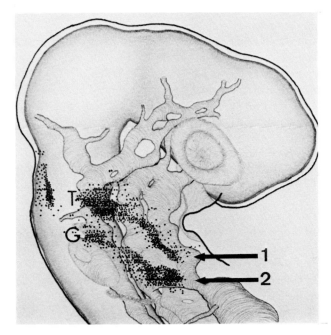

Fig. 1.6 Similar to Fig. 1.5 except that the migrations of anterior hindbrain neural crest cells are shown. Crest cells are located in the primordia of the trigeminal (T) and geniculate (G) cranial sensory ganglia, and in the first (1) and second (2) visceral arches. (From Noden, 1975).

mesenchymal wedge which fills the region between the lateral margin of the hindbrain and the superficial ectoderm. Most of these cells then move ventrally between the epidermis and the mesoderm, while the rest remain adjacent to the ventro-lateral aspect of the hindbrain, immediately dorsal to the anterior cardinal vein (Noden, 1975).

The rhombencephalic neural crest population becomes subdivided segmentally during this early migratory phase. As shown in Fig. 1.6 this results in the appearance of aggregations of crest cells adjacent to the future loci of trigeminal and facial motor nerve emergence. These aggregates are later joined by cells emigrating from superficial ectodermal placodes, and together they give rise to cranial sensory ganglia (van Campenhout, 1937; Yntema, 1944; Hamburger, 1961; Johnston, 1966; Noden, 1975). The large, ventrally moving masses of neural crest cells, which are also segmented (Fig. 1.6), migrate past the lateral margin of the pharynx and populate the first (mandibular) and second (hyoid) visceral arches. At a later stage, these connective tissue-forming masses lose contact with the proximal, presumptive ganglionic condensations. The migratory patterns of post-otic rhombencephalic neural crest cells are similar to those just described, although the environment through which they migrate more closely resembles that found in the trunk. The initial size of the ventrally moving mass is smaller than was observed in pre-otic regions. Also, in addition to forming visceral arch connective tissues, some of these crest cells migrate caudally, apparently using the endodermal epithelium as a substratum, and form most of the enteric parasympathetic neurons (Le Douarin and Teillet, 1971, 1973; see Section 1.3.3).

At the time of diencephalic neural crest formation the body wall ectoderm is tightly apposed to the prosencephalon and the evaginating optic vesicles. Only the caudal (future dorso-temporal) surface of the eye is exposed, and it is associated with a sparse population of mesodermal mesenchyme. Crest cells from this region, together with anterior mesencephalic crest cells which have migrated rostrally (Noden, 1975), initially mass above the constricting optic stalks. As constriction continues some of these crest cells migrate over the caudal and, later, rostral surfaces of the optic stalk and vesicle.

Many crest cells remain dorsal to the prosencephalon and eyes. Subsequent rostral movements and proliferative expansion of this population results in the formation of the frontal prominence and nasal processes (Johnston, 1966; Johnston and Hazelton, 1972; Johnston, 1974), which later fuse with the maxillary process during facial morphogenesis. Cells which move laterally around the optic vesicle later invade the region between the lens and the superficial ectoderm, forming the corneal endothelial and stromal layers and contributing to the iris (Fig. 1.10; see Section 1.3.1).

1.2.3 Morphology of migrating neural crest cells

During the migratory phase of their development, many neural crest cells are flattened and polarized with their long axes parallel to the direction of movement. The distal part of the cell is enlarged and, as shown in Fig. 1.7, the leading edge forms a lamellipodium-like process. The Golgi apparatus is located in the tapering, trailing region of the cell. Thus, these crest cells are morphologically similar to migrating sclerotomal (Trelstad *et al.*, 1967; Trelstad, 1977; Ebendal, 1977) and corneal stromal (Bard and Hay, 1975) cells *in vivo*. This bipolar, spindle-shaped morphology is observed *in vitro* when neural crest cells are grown in three-dimensional collagenous lattices (Bard and Hay, 1975; Davis, 1977), but is lost when they are cultured on flat, two-dimensional substrate (reviewed by Trinkaus, 1976).

Neural crest cells also numerous filopodia projecting from their surface. Many of these long, slender processes appear to contact extracellular fibers and basal laminae of adjacent tissue (Bancroft and Bellairs, 1976; Tosney, 1978). In addition, these filopodia frequently form plaque contacts with other crest cells (Johnston and Listgarten, 1972; Ebendal, 1977; Steffek *et al.*, in preparation). Such contacts are similar to the transient junctions seen between colliding fibroblasts *in vitro* (Heaysman and Pegrum, 1973).

The plasma membranes of migrating crest cells are frequently closely apposed (20 nm intercellular space, Ebendal, 1977) over large areas. While such regions are generally free of other surface specializations, discrete patches containing

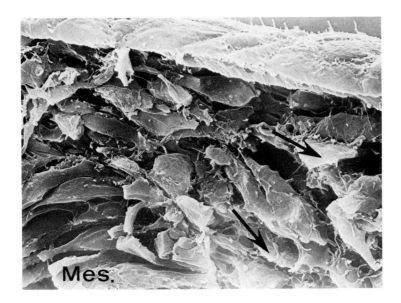

Fig. 1.7 Scanning EM of migrating neural crest cells located dorso-lateral to the mesencephalon (Mes.). Arrows indicate the direction in which the crest population is migrating. (Courtesy of K. Tosney).

electron-dense material may be found. These patches are morphologically similar to the plaque contacts described above. Tight junctions, which are formed between migrating mesoderm cells during gastrulation (Hay, 1968) have not been observed between crest cells.

1.2.4 The neural crest environment

Interactions between embryonic cells and their environment play an obligatory role in most morphogenetic and differentiative processes (reviewed by Wessells, 1977). Their nomadic behavior exposes migrating neural crest cells to a complex, heterogeneous array of cellular and extracellular components. In order to assess the role of such influences in directing crest cell migrations and subsequent cytodifferentiation, it is necessary to identify and define these components and to characterize their spatial and temporal distribution.

(a) *Cellular and cell-bound components*
The cell type with which migrating neural crest cells are in most frequent and direct contact is other neural crest cells. While the precise function of the plaque junctions and close appositions described in the preceding section is unknown, maintaining the integrity of the mesenchymal population is an essential pre-requisite for normal later development (see Section 1.5.2 (a)).

Some crest cells are seen in contact with the basal surface of either the superficial or the neural epithelium. Migrating crest cells do not touch the plasmalemma of these epithelia directly; rather, they are associated with components of the basal lamina or, when present, the extra-laminar (ground substance, fibrous sheath) meshwork. These layers are difficult to characterize precisely because neither the precursor pool sizes nor the rates of synthesis and degradation of many of their components are known. In addition, it is difficult to separate permanent laminar-associated materials from those which are being released into the extracellular spaces (reviewed by Manasek, 1975).

Immediately prior to neural fold fusion the ectoderm is underlain by a continuous basal lamina which is rich in both sulfated and non-sulfated glycosaminoglycans (Fisher and Solursh, 1977). As fusion occurs the lamina is disrupted within the folds (Meade and Norr, 1977). The data of Bancroft and Bellairs (1975) indicate that a continuous basal lamina is re-established both beneath the superficial ectoderm and over the roof of the brain prior to cephalic neural crest emigration. In contrast, Tosney's (1978) observations on neurulation in the trunk show that the lamina remains discontinuous over the roof of the spinal cord until after crest cell migration is under way. The appearance of interruptions in the basal lamina is thus correlated both spatially and temporally with the initial formation of the neural crest population in these different regions.

The material found on the lateral and ventral margins of the spinal cord has been characterized largely through the efforts of Hay and her colleagues. At the time of crest cell migration the neuro-epithelial basal lamina is a 300 nm thick, electron-dense layer (Cohen and Hay, 1971). Ruthenium Red binding and isotope incorporation studies indicate the presence of both sulfated and non-sulfated glycosaminoglycans (Hay and Meier, 1974), and an increase in the amount to ^{3}H-proline-containing material.

The overlying material, which is also secreted by the neural epithelium, contains a meshwork of fibrils, most of which have a diameter of 200 nm. While banded fibrils are not seen until later stages of development, the major fibrous component within the meshwork appears to be collagen. This layer is also rich in glycosaminoglycans. Manasek (1975) reports an increase in the amount of ^{3}H-glucosamine-incorporating material adjacent to the neural tube, especially the ventral region, at the time of neural crest emigration. The most abundant glycosaminoglycan within this layer is hyaluronic acid (Manasek and Cohen, 1977). Sulfated material is present but is being released into the extracellular space.

The basal surface of avian cephalic superficial ectoderm has been examined by Fisher and Solursh (1977) using the scanning electron microscope. Immediately prior to neural crest emigration this epithelium has a basal lamina which is covered with a fibrous material sufficiently dense to occlude the outlines of underlying cells. Individual filaments are coated with a fuzzy granular material which is hyaluronidase-sensitive. Steffek *et al.* (in preparation) have found some fibers which are aligned transverse to the body axis, and thus are parallel to the direction of subsequent neural crest cell movement. However, most of the fibers associated with both the superficial and the neural epithelia are randomly oriented in birds.

Glycoproteins are also present in the basal laminae of these embryonic epithelia, as indicated by the presence of ^{3}H-fucose-labeled material (Manasek, 1975). However, characterization of these components and analysis of their functional importance are just beginning (Manasek and Cohen, 1977).

(b) *Extracellular matrix components*
Neural crest cells initially move between rather than into adjacent embryonic tissues. Histological examination of avian embryos shows that large separations between tissues appear prior to the movement of crest cells into a region. This is most apparent in the head, as illustrated in Fig. 1.8.

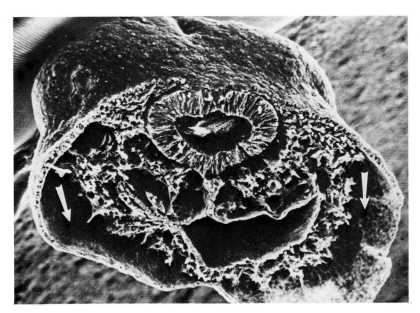

Fig. 1.8 Scanning EM of a chick embryo broken in a transverse plane through the posterior mesencephalon. The neural folds have fused and crest cells are seen lateral to the brain. The arrows indicate large cell-free spaces which appear between the superficial ectoderm and the mesoderm ahead of the advancing crest population. (Courtesy of K. Tosney).

Scanning electron microscopic analysis of these extracellular spaces reveals a rich fibrous meshwork composed of randomly arranged 50—100 nm fibers, many of which are continuous with extra-laminar material (Bancroft and Bellairs, 1976; Tosney, 1978). The presence of periodic cross-banding suggests that they are collagen. When cetylpyridinium chloride, which precipitates glycosaminoglycans, is included in the

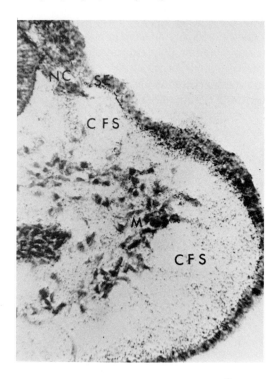

Fig. 1.9 Autoradiagram of a chick embryo treated with ^{3}H-glucosamine
3 hours prior to fixation. Glucosamine products are found in the cell-free
space (CFS) into which crest cells (NC) are migrating. (With permission from
Pratt *et al.,* (1975) and Academic Press.)

fixative the appearance of the meshwork is altered dramatically. The filaments are
now 200–300 nm in diameter due to the presence of granular, often beaded material
which largely obscures the native fiber (Fisher and Solursh, 1977; Tosney, 1978).
With the onset of crest cell migration the number of filaments increases dramatically,
and is greatest as the cells move through the former cell-free spaces.

Analyses using ^{3}H-glucosamine labeling (Fig. 1.9; Pratt *et al.,* 1975), enzyme
treatment (Pratt *et al.,* 1976; and alcian blue histochemistry (Derby *et al.,* 1976)
all indicate that hyaluronic acid is the major glycosaminoglycan component in these
spaces, although sulfated material can be detected. Hyaluronic acid is a high
molecular weight anionic mucopolysaccharide that is unusual in that when released
by cells it can occupy a molecular domain 10 000 times greater than that of other
compounds of similar molecular weight. The appearance of this glycosaminoglycan
has been correlated with many morphogenetic events, which have been reviewed
by Toole (1976). For example, it is an integral component in heart (Manasek, 1976),
limb (Toole, 1972) and cornea (Toole and Trelstad, 1971; see Section 1.3.1)

development, and is also associated with cell migrations during avian gastrulation (Solursh, 1976).

In the head region, there are two major sources of hyaluronic acid at the time of extracellular space formation. Pratt *et al.* (1976) have shown that isolated superficial ectoderm will incorporate ^{3}H-glucosamine into hyaluronic acid, most of which is released. In addition, cranial neural crest cells *in vitro* release glycosaminoglycan, most of which is hyaluronic acid (Greenberg and Pratt, 1977).

The situation is more complicated in the trunk. The neural tube, somites, and trunk neural crest cells have all been shown to synthesize both sulfated and non-sulfated glycosaminoglycans. However, the relative contributions of these tissues to the extracellular material has not been defined.

Also found within the fibrillar meshwork are electron-dense, granular bodies of $0.1-1.0\ \mu m$ diameter (Tosney, 1978; Low, 1970). These bodies, whose function is unknown, are most abundant near the leading edge of the migrating neural crest population and are also seen associated with basal laminar fibrils adjacent to the neural tube (Cohen and Hay, 1971).

This characterization of a migrating neural crest cell's environment is far from complete. Yet to be defined are the types of collagen present in both cell-associated and extracellular matrices, the precise ratios and distribution of sulfated and non-sulfated glycosaminoglycans, the nature and role of cell-surface and secreted glycopeptides and, most importantly, the ways in which these components interact with one another. As will be discussed in Section 1.5 and 1.6, these environmental components perform essential functions in directing the migration and cytodifferentiation of neural crest cells.

1.3 LATER MOVEMENTS OF NEURAL CREST CELLS

1.3.1 Formation of the cornea

While the events just described bring most neural crest cells to their terminal locations, some crest cells display additional migrations. The best characterized of these occur during the formation of the avian cornea (reviewed by Hay and Revel, 1969). The flattened, relatively isolated nature of this tissue renders it especially suitable for cytological and biochemical analyses of migratory processes both *in situ* and *in vitro*.

Following lens formation, a narrow space, the future anterior chamber, exists between the lens and the superficial ectoderm, now called the anterior corneal epithelium. During the third and fourth days of incubation this epithelium produces the primary stroma, a highly convoluted lattice of orthogonally arranged layers of 25 nm-diameter collagen fibers (Hay and Revel, 1969; Trelstad and Coulombre, 1971; Dodson and Hay, 1971). By the end of the fourth day of development the primary stroma is $5-10\ \mu m$ thick and contains approximately 20 layers.

The anterior chamber is bound circumferentially by a fibrous meshwork connecting the edge of the optic cup with the stroma (Bard *et al.,* 1975). Outside of this meshwork is the richly vascularized periocular mesenchyme. The only cells from this tissue to enter the cornea during the initial 3½ days of incubation are macrophages, which are found between the stroma and the lens.

Near the end of the fourth day of development the circumferential fibrous mesh-work swells and mesenchymal cells move in. These cells, which contain an abundance of endoplasmic reticulum and Golgi material, continue migrating and enter the future anterior chamber. They do not invade the primary stroma, but rather migrate between it and the lens capsule to form the corneal endothelium.

While a few cells appear to be migrating alone, most are in small clusters or are part of a loose monolayer (Nelson and Revel, 1975). Generally, the migrating corneal endothelial cells are flattened and elongated, although there is no consistent orientation to their long axes. Most have a lamellipodium similar to that seen on early migrating crest cells and several filapodia which contact the primary stroma the lens capsule, or other nearby cells.

Hay and Revel (1969) and Bard *et al.* (1975) have suggested that the corneal endothelium is formed by mesodermal cells, possibly having the same embryonic origin as periocular vascular endothelium. The basis of this assertion is Hay's (1968) classification of the neural crest as a component of secondary mesenchyme, which normally does not give rise to epithelial tissues. However, transplants of quail neural crest cells into chick embryos have clearly shown that the corneal endothelium, but not adjacent vascular endothelium, is entirely of neural crest origin (Fig. 1.10; Johnston, 1974; Noden, 1976). Quail cells have a dense condensation of nucleolar-associated heterochromatin not found in chick nuclei (Le Douarin 1969, 1971). Because it is self-replicating and visible in all differentiated quail cell types, the marker can be used to follow later aspects of neural crest development.

During the fifth day of incubation the acellular primary stroma doubles in width and, towards the end of this period, is invaded by periocular mesenchymal fibroblasts. These cells initially invade the stroma adjacent to the endothelium, but soon occupy the entire region, except for a narrow zone adjacent to the anterior epithelium which will become Bowman's membrane (Fig. 1.10). The presence of extensive endoplasmic reticulum and Golgi complexes indicates that these fibroblasts are actively secretory (Hay and Revel, 1969), and several studies have shown them to be the source of the collagen(s) which are present in the mature cornea (Conrad, 1970; Trelstad and Coulombre, 1971). By 10 days of development the stroma reaches its maximum width, slightly over 200 μm. Subsequent dehydration results in reduction of corneal width and the appearance of transparency (reviewed by Coulombre, 1965).

Little is known about the non-collagenous extracellular components of the cornea during endothelium formation. On the fourth day of development the anterior epithelium produces equal amounts of heparin sulfate and chondroitin sulfate (Meier and Hay, 1974), but no unsulfated glycosaminoglycans. In contrast, the swelling of the primary stroma on the fifth day of incubation is correlated with a dramatic

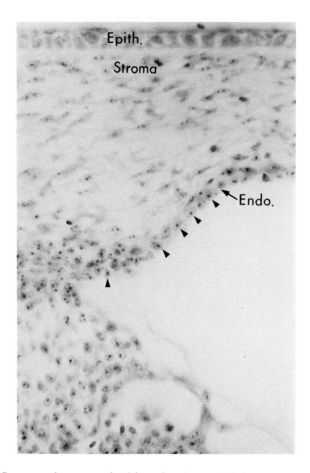

Fig. 1.10 Cornea and presumptive iris region from a 7½ day chick embryo whose mesencephalic neural crest population was replaced with head crest tissue excised from a quail embryo. Quail crest cells which have contributed to the corneal endothelium (Endo.) are easily identified by the presence of a mass of nucleolar-associated heterochromatin (indicated by arrowheads). Most cells in the stroma and presumptive iris regions also contain this marker. Cells in the anterior corneal epithelium (Epith.) are not of neural crest origin and thus do not contain this marker.

increase in the amount of hyaluronic acid (Toole and Trelstad, 1971). Meier and Hay (1974) suggest that the endothelium must be the major initial source of hyaluronic acid, with corneal fibroblasts themselves subsequently producing it (Dahl *et al.,* 1974). However, the relative contribution of these two tissues to the extracellular milieu is not known.

1.3.2 Movements along nerve fibers

A neural crest origin for Schwann sheath cells was first proved by Harrison (1906) in a pioneering series of extirpation experiments on amphibian embryos. This has been confirmed in birds by Weston (1963), Johnston (1966) and Noden (1975), although these authors recognize that some sheath cells emigrate from the ventral neural tube along ventral roots (Harrison, 1924; Triplett, 1958; Jaros and Noden, in preparation). Fig. 1.11 clearly shows the extent to which crest cells utilize nerve fibers as a substratum for migration. In this case, transplanted, ^{3}H-thymidine labeled crest cells migrated in

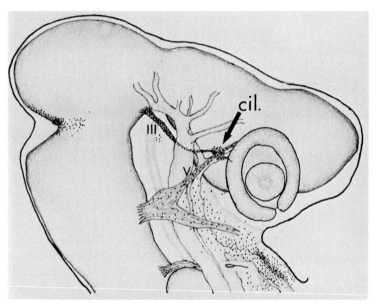

Fig. 1.11 Reconstruction of a 3-day chick embryo as in Figs. 1.5 and 1.6. While most cells from the mesencephalic-metencephalic junction migrated into the mandibular arch, a few moved towards the eye, perhaps using the ophthalmic nerve as a guide. Some of these crest cells are seen clustered where this nerve crosses the path of the oculomotor nerve (III). This aggregate is the primordium of the ciliary parasympathetic ganglion. Other crest cells are found distributed along the length of the oculomotor nerve. (From Noden, 1975.)

a distal-to-proximal direction along the oculomotor nerve. The absence of labeled cells in most adjacent tissues suggests the crest/neurite association is strong.

Direct observations of cell movement along nerve fibers in the transparent tail fin of the tadpole have been made by Speidel (1932) and Billings-Gagliardi *et al.* (1974). While moving, the cells are pleotropic and have lamellipodia and filopodia. Migration occurs erratically both along the fibers and in the adjacent mesenchyme. Cessation of movement is correlated with elongation of the presumptive Schwann cell and with the formation of first a patchy then a continuous basal lamina (Billings-Gagliardi *et al.,* 1974), a process which is reversible both during normal development and following injury (see Abercrombie *et al.,* 1949). Although known to contain collagen (Church *et al.,* 1973), neither the composition nor the role of this basal lamina during active presumptive Schwann cell migration are understood.

Some evidence suggests that not all cells moving along nerve fibers are committed to become Schwann cells. For example, the cluster of cells present at the site of oculomotor and ophthalmic nerve crossing in Fig. 1.11 is the presumptive ciliary ganglion. A probable, but not proved, pathway along which these crest cells migrated is the ophthalmic nerve, although it is possible that both the pioneering ophthalmic nerve fibers and the presumptive ciliary neurons and oculomotor sheath cells moved into this periocular region simultaneously. It has also been suggested that presumptive intrinsic visceral parasympathetic neurons migrate along descending vagus nerve fibers (Yntema and Hammond, 1945).

These experiments indicate that neural crest cells use developing nerve fibers as substrata for migration. In addition, the *in vitro* analyses of Wood and Bunge (1975) reveal that neurites have a mitogenic effect on Schwann cells. The possibility that nerve fibers might affect the proliferation and, perhaps, the determination of some neural crest cells remains to be investigated.

1.3.3 Migration of presumptive enteric parasympathetic neurons

That neural crest cells emanating from the myelencephalon are the major source of enteric parasympathetic neurons, including Meissner's and Auerbach's plexes, has been confirmed by extirpation (Yntema and Hammond, 1954) and explantation (Andrew 1970, 1971) studies. Le Douarin and Teillet (1973) have followed the movements of labeled vagal crest cells following the surgical replacement of the chick myelencephalon with a homologous piece of quail tissue. These authors found quail cells scattered in the splanchnic mesenchyme surrounding the esophagus by 3 days of incubation (approximately 36 hours after transplantation), at the umbilicus by 5 days and the rectum by 8 days. Unfortunately, there have been no studies analyzing the structure of these cells, their immediate environment, or the mechanisms underlying this remarkably long migratory phenomenon.

1.4 AVIAN NEURAL CREST DERIVATIVES

Classical descriptive and experimental embryological studies have demonstrated that the neural crest is the source of all autonomic neurons, most peripheral sensory neurons, Schwann sheath cells and supporting cells of the peripheral nervous system, all pigment cells except those of the neural retina and brain, and the splanchnocranium and rostral part of the chondrocranium (Table 1.1; also reviewed by Hörstadius, 1950;

Chibon, 1966; Weston, 1970). However, using these techniques it was not possible to detect minor or late-forming crest derivatives nor to distinguish crest-dependent from crest-derived structures (discussed by Andrew, 1971). These problems have been solved by surgically creating quail neural crest: chick host chimeras, as discussed in Section 1.3.1. This method has greatly expanded the number of connective tissues and secretory cells known to be of neural crest origin.

As listed in Table 1.1, most of the skeletal and connective tissues of the face, palate, mouth and jaw are derived from the cranial neural crest (Le Lièvre, 1974; Le Lièvre and Le Douarin, 1975). In contrast, the trunk neural crest (caudal to the sixth somite) does not contribute to any skeletal or connective tissues in birds.

Pearse (1969) suggested that all polypeptide-secreting endocrine cells that are capable of sequestering and decarboxylating amine precursors such as dihydroxy-phenylalanine (DOPA) might share a common embryonic origin. Included in this series are most endocrine cells associated with gut (endodermal) epithelium and its derivatives, such as the pancreas and lungs, and also some adenohypophyseal cells. While neural crest cells form several of these polypeptide-secreting cells, including the adrenal medulla, calcitonin-producing cells (Le Douarin and Le Lièvre, 1970) and carotid body type I cells (Pearse *et al.*, 1973), recent studies indicate that some of the secretory cells in Pearse's series are of endodermal not neural crest origin (Andrew, 1976; Picted *et al.*, 1976; Fontaine *et al.*, 1977).

1.5 SPECIFICATION OF MIGRATORY PATHWAYS

Morphogenetic movements are common events during early stages of animal development (reviewed by Trinkaus, 1969; Sidman, 1974), yet the mechanisms controlling most of these processes are poorly understood. Clearly, the temporally and spatially patterned movements of avian neural crest cells are the result of inter-actions between each crest cell and components of its environment. Despite their complexity, neural crest cells offer two important advantages to investigators of vertebrate morphogenesis. First, these cells are readily accessible to biochemical and surgical manipulation during both the initial and certain later stages of migration. In addition, regional differences in migratory patterns and composition of the environ-ment enable one to use a comparative approach to investigate the mechanisms underlying patterned cell movements.

1.5.1 Control of migratory behavior

Understanding the control of neural crest cell migration is dependent in part upon defining the relative role(s) that each component of the intra-embryonic environment plays in influencing crest cell behavior. These will be discussed in Section 1.5.2.

Equally essential, but more difficult to approach, is the response of an individual crest cell to environmental factors. Since the pathways of neural crest cell migration differ even among those cells emigrating from the same region, it is possible that the initial crest population is in fact heterogeneous. Within the pre-migratory crest tissue might exist subpopulations each capable of responding differently to common extrinsic environmental influences.

In vitro analyses have shown that the behavior of crest cells isolated from the cornea is different than fibroblasts of heart or trunk dermal origin (Conrad *et al.,* 1977). However, avian pre-migratory neural crest cells explanted onto culture dishes all initially have a uniform, stellate morphology and exhibit a similar behavior (Cohen and Konigsberg, 1975; Maxwell, 1976; Greenberg and Schrier, 1977). Important differences between crest cells could exist, of course, and not be expressed under these restrictive culture conditions. Because the identity and distribution of those cell surface and intracellular components which facilitate the response of a crest cell to an environmental stimulus are as yet poorly defined (Manasek and Cohen, 1977), no direct biochemical analysis of individual crest cell properties is currently feasible.

Analyses of the migratory potentialities of neural crest *populations* have been done by exploiting the regional differences in migratory behavior outlined in Section 1.2.2. Noden (1973, 1975) replaced chick mesencephalic crest cells, which normally disperse uniformly (Fig. 1.5), with ^{3}H-thymidine-labeled or quail metencephalic crest cells, which coalesce to form cranial ganglia and also invade the visceral arches (Fig. 1.6). In these cases, the hindbrain-derived crest cells dispersed, thus behaving in a manner appropriate to their new environment. Not only are the initial migratory patterns implantation-site-specific, but later movements such as invasion of the cornea also occur normally. The cornea shown in Fig. 1.11 is in fact composed of neural crest cells which migrated from a population in hindbrain crest cells implanted in the place of the anterior midbrain neural crest.

Reciprocal transplantations give identical results. Mesencephalic and diencephalic crest cells, which would normally not form presumptive ganglionic aggregations, will do so if they are allowed to migrate beside the hindbrain. In these cases, development of the visceral arches is also normal (Noden, 1978a).

Exchanges between cephalic and trunk neural crest cells have also been made. Noden (1975) replaced the metencephalic crest primordium with labeled dorsal spinal cord fragments, which had been treated with trypsin to remove any adhering somitic cells. As shown in Fig. 1.12, these transplanted crest cells migrate in a manner indistinguishable from that exhibited by metencephalic crest cells. Le Douarin and Teillet (1974) replaced the entire post-otic myelencephalon of a chick embryo with an equivalent length of quail thoracic spinal cord. The grafted trunk quail cells mimic the behavior of hindbrain crest cells by migrating caudally adjacent to the gut epithelium to form enteric parasympathetic plexes. This represents a migratory behavior which thoracic neural crest cells would not normally display.

The reciprocal grafts in part yielded similar results. Crest cells emigrating from segments of the rhombencephalon, which had been implanted in the place of the thoracic spinal cord, migrated to the location of spinal sensory ganglia, para- and pre-vertebral sympathetic ganglia, the supra-renal glands (diffuse adreno-medullary

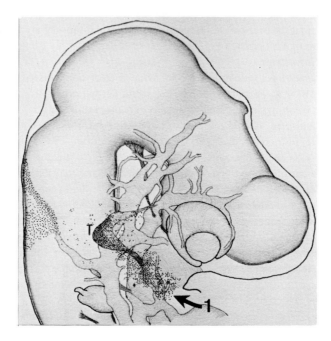

Fig. 1.12 Reconstruction of an 80-hour chick embryo whose metencephalic crest cells were replaced with ³H-thymidine-labeled brachial neural crest cells. These transplanted crest cells have migrated into the first visceral arch (1) and contributed to the formation of the trigeminal ganglion (T) in a manner identical to normal metencephalic crest cells (compare with Fig. 1.6). (From Noden, 1975.)

tissues), and pigment cells of the integument. However, some interesting departures from normal thoracic patterns of crest cell migration were also observed. Quail crest cells invaded and subsequently contributed to connective tissue derivatives of the adjacent sclerotomes and dermatomes. Most surprisingly, crest cells from the transplant migrated to the intestinal wall where they contributed to enteric ganglia. Thus, some crest cells migrated in a manner appropriate to their original embryonic location, but inappropriate for their site of implantation.

These results may be due simply to differences in initial population size and rate of emigration, both of which are greater in the cranial regions. However, the experiments of Le Douarin and Teillet raise the possibility that there are subsets of cells within the cranial neural crest population which differ in their migratory potentiality. For example, one group of cranial crest cells, those which normally form connective tissues, might invade and remain interspersed with somitic cells. Possibly some hindbrain crest cells recognize and are responsive to environmental influences permitting (promoting?) migration ventrally towards the gut. However, regardless of their origin, once neural crest cells become associated with the mesenchyme surrounding the gut they behave in an identical manner, migrating caudally to form enteric parasympathetic neurons.

Since the environments which must be traversed to reach the gut from the hindbrain and from the mid-spinal regions are morphologically dissimilar, Le Douarin and Teillet (1974) have suggested that the gut releases an attractant to which only hindbrain crest cells are responsive. As with similar proposals for guidance of nerve outgrowth (Chamley *et al.*, 1973, Chamley and Dowel, 1975; Coughlin, 1975; Pollack and Liebig, 1977), such chemotropic mechanisms are difficult to verify *in situ*. Epperlein (1974) has examined the behavior of amphibian cranial neural crest cells alone or in combination with fragments of the pharyngeal endoderm *in vitro*, and is unable to detect any oriented cell movements until the crest cells are within 50 μm of the epithelial tissue.

Evidence for an inherent, region-specific migratory behavior was also presented by Chibon (1966) who, working with a salamander embryo, transplanted ^{3}H-thymidine-labeled hindbrain neural folds in the place of forebrain folds. He found that crest cells from the graft contributed to the mandibular teeth, which are normally of hindbrain crest origin, and not to the palatine teeth, which are of mid- and fore-brain crest derivation.

Thus, the transplantation experiments do not unequivocally define the migratory potentialities of pre-migratory neural crest cells. Clearly the specific behavior of every crest cell is not programmed at the onset of emigration. For many of these cells the migratory pathways taken and associations made are the result of interactions between the crest cells and their environment. On the other hand, the possibility that some crest cells within a population possess regionally-specific migratory capabilities cannot be excluded. In addition, while these experiments do indicate that the migratory potentialities of transplanted neural crest *populations* are very similar, they do not prove that each individual crest cell within the population is equipotential.

Weston (1963) observed that the first neural crest cells to emigrate from the chick spinal cord migrate the furthest and contribute to the most distant (ventral) structures, while the last to leave normally participate in sensory ganglion formation near their origin. This suggests that the early migrators might be different from the later-formed crest cells in their migratory abilities. To test this Weston and Butler (1966) replaced a segment of spinal cord, from which neural crest migration had not begun, with a ^{3}H-thymidine-labeled segment from which the early-formed crest cells had already emigrated. The result was that cells from the graft migrated to all of the locations normally occupied by trunk crest; the last-formed neural crest cells have the same migratory capabilities as the first. Thus, if there are unique subsets of cells within the crest population, they are not segregated according to the time or the order of neural crest formation. This also appears to be true in the head regions, in which some early emigrating crest cells contribute to the same structures as those formed later (Noden, 1975).

1.5.2 Role of specific environmental components

The experiments discussed above indicate that the diverse patterns of neural crest cell movement and distribution are the result of interactions between each crest cell and the intra-embryonic milieu. Most probably the migratory behavior of these cells is an integrated response to the temporal and spatial arrangements of many different environmental components. The aim of this section is to analyze the role of some of these components in directing the migration of crest cells.

(a) *Other neural crest cells*

During the early stages of migration the cell type with which each neural crest cell is most frequently in contact is another crest cell. The importance of these contacts in establishing (or maintaining) migratory patterns has been demonstrated by experiments in which the crest population is surgically disrupted. Following the removal of a short segment of pre-migratory neural crest cells, crest cells emigrating from sites adjacent to the gap deviate from their normal routes and invade the now unoccupied regions (see, for example, Yntema and Hammond, 1945; Hammond and Yntema, 1947, 1958, 1964; Nawar, 1956). In addition to altered patterns of migration, these crest populations undergo compensatory hyperplasia, usually resulting in the formation of structures which are of normal size and morphology. It is this regulative capacity which makes the interpretation of neural crest ablation experiments difficult, as discussed by Weston (1970) and Andrew (1971).

These experiments prove that crest cells are not simply following unalterable 'tracks' towards their destinations. Moreover, they indicate than an essential prerequisite for the expression of normal migratory patterns is the integrity of the entire neural crest mesenchymal population.

Possibly, the maintenance of an intact population is simply a *sine qua non* for the movement of crest cells away from their origin and is not necessary for the establishment of heterogeneous migratory patterns. Twitty proposed that the early dispersal of neural crest cells might be the result of a chemically mediated mutual repulsion (Twitty, 1949; Twitty and Niu, 1948, 1954). This hypothesis, which was based on the behavior of amphibian melanophores *in vitro*, appears incompatible with the more recent observations that those cells which break away from the leading edge of the crest population either *in vitro* or *in vivo* usually re-establish close contacts.

Based on *in vitro* analyses of fibroblast cell movement, Abercrombie proposed that mesenchymal cell behavior could best be explained by a 'contact inhibition of movement' mechanism (reviewed by Abercrombie, 1970). Briefly stated, this model postulates that when active migration brings the leading process of a cell in contact with the surface of another cell, the first cell will stop moving, form a new lamellipodium and move in a different direction if possible. In the case of a mesenchymal population, for example the neural crest *in vivo*, contact inhibition of movement would result in the migration of cells away from areas of high cell density and into available cell-sparse regions such as the cell-free spaces described in Section 1.2.4(b), the acellular matrix between the anterior corneal epithelium and lens (Section 1.3.1), or the crest cell-free regions resulting from neural crest ablation.

(b) *Possible barriers to cell movement*
Both the timing and the pattern of migration of some crest cells may be affected by environments which block cell movement. Figure 1.13 shows neural crest cells dorsal to the myelencephalon at the region of otic placode invagination. Transplantation studies (Noden, 1973, 1975) indicate that these crest cells will move rostrally or

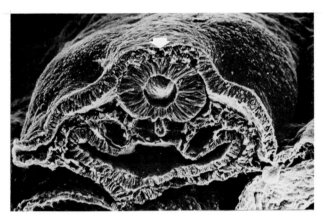

Fig. 1.13 Scanning EM of a chick embryo broken in a transverse plane near the invaginating otic placodes. Crest cells which form in this region (arrow) migrate rostral or caudal to the otic placode rather than in a ventro-lateral direction beneath it, suggesting that the placode may act as a physical barrier. (Courtesy of K. Tosney.)

caudally around the otic region before migrating ventrally towards the visceral arches. Whether the inability of these crest cells to initially migrate ventrally beside the hindbrain is a result of occlusion of the extracellular space, as Figure 1.13 might suggest, or an inability of the milieu beneath the otic epithelium to support cell migration (see next section) has not been established.

Noden (1975) reported that some anterior mesencephalic crest cells move rostrally in the dorsal midline and join with cells of prosencephalic origin above the optic stalks. Figure 1.1 indicates that the dorso-lateral margin of the midbrain is contacted by thickenings of the superficial epidermis at the time of crest cell formation. That these areas of apparent contact are in fact zones of tight adherence has been demonstrated by the appearance of a distortion in the wall of the neuro-epithelium when the outer ectodermal layer is surgically lifted. Such barriers might serve to delay the lateral movement of mesencephalic crest cells and thereby promote a rostral migration. Bard *et al.* (1975) have noted that when a similar type of attachment between the lens and the lip of the optic cup is inadvertently broken, some crest cells deviate from their normal intra-corneal migratory route and move along the presumptive neural retina within the optic cup.

Another proposed barrier to cell migration is the dense fibrous meshwork which exists around the presumptive cornea between the lip of the optic vesicle and the anterior corneal epithelium. Prior to the movement of prospective endothelial cells into the cornea, macrophages enter the region and swelling of the matrix occurs (Coulombre *et al.,* in preparation). While these events may be co-incidental, it is tempting to suggest that macrophage activity and/or a hyaluronate-mediated matrix swelling act to open the future cornea to neural crest invasion. The final distribution of corneal endothelial cells is dependent upon the presence of both the lens and the anterior epithelium. If the lens is removed, the neural crest population entering the region below the primary stroma maintains a multilayered, mesenchymal appearance rather than forming a monolayer (Bard *et al.,* 1975). In this case, the lens acts to confine the crest population within a single plane.

These experiments indicate that the timing of migration and spatial distribution of neural crest cells may in some instances be affected by the presence of barriers which prevent cell movement in a certain direction. However, many patterned migratory phenomena occur independent of the formation or removal of environmental barricades. In addition, as stated by Weston (1963, p. 301), 'For every instance where lodging appears to be the result of blocked migration, there are instances of cells avoiding the same barrier'.

(c) *Permissive environments and supportive substrata*
Ideally, those environmental components which actively promote and direct neural crest cell migration should be analysed and discussed separately from those whose function is to allow for expansion of the crest population in a given direction. In some situations, such as the movement of presumptive Schwann cells along nerve fibers or of enteric parasympathetic neurons in the splanchnopleure of the gut, this distinction can be made. However, in many cases it is not possible to distinguish permissive from directive influences. This is because both the spatial distribution and the functional properties of many extracellular components are dependent upon the composition of the milieu in which they are located (Toole, 1976). Thus, it is difficult to analyze experimentally the role of one component without at the same time disrupting the entire environment. This, together with our inability to exclude the possibility of differential responsiveness (i.e. heterogeneity) on the part of neural crest cells (Section 1.5.1), means that assigning a specific role to any component in the neural crest environment must be done cautiously and, usually, tentatively.

The presence of newly synthesized hyaluronic acid and formation of a cell-free space (or swollen matrix) frequently occur concomitant with or in advance of cell migration during vertebrate embryogenesis, suggesting that these events might be causally related. Its appearance in the head regions of the chick is temporally correlated with a loss of contacts between mesoderm and the superficial ectoderm, which may have been restricting neural crest movement. Application of *Streptomyces* hyaluronidase to avian embryos at gastrula and neurula stages results in the loss of cell-free spaces, a reduction in the amount of extracellular material, and the clumping of mesodermal mesenchyme (Fisher and Solursh, 1977). Under these conditions the emigration of neural crest cells is prevented (Pratt *et al.,*1976).

One explanation of the mode of action of hyaluronic acid in early crest cell movement and corneal fibrobalst invasion is that its presence provides a suitable milieu for cell migration (Toole, 1976), i.e. acts in a permissive manner. The possibility that this glycosaminoglycan interacts with other environmental components, perhaps altering their suitability as substrata for cell attachment, cannot be excluded. However, there is no evidence which indicates that the heterogeneity of migratory behaviors exhibited, for example, by hindbrain and trunk-derived crest cells can be attributed to patterns of hyaluronic acid synthesis or degradation.

Weston (1963, 1971) has suggested that crest cell migration is facilitated by 'organized cellular environments' such as the somitic mesenchyme. This hypothesis is based on his observation that transplanted trunk crest cells move within but not between somites, regardless of the direction from which they enter. These data confirm previous analyses of amphibian crest cell migration by Lehmann (1927) and Detwiller (1934). Subsequently, the somites lose their ability to support the migration of crest cells (Weston and Butler, 1966), a change which is correlated with both a decrease in the level of somitic hyaluronic acid (Toole, 1972) and, in the sclerotome regions, an increase in the level of sulfated glycosaminoglycans (Derby *et al.,* 1976). These alterations in the somitic matrix composition might be responsible for permitting early emigrating crest cells to traverse the somites while, at later stages, crest cells are restrained and subsequently form dorsal root ganglia.

While this hypothesis is plausible, any proposed effects of the somitic environment on crest cell migration must be compatible with the results of the heterotopic, trunk-into-head neural crest transplants discussed previously. In those experiments (Noden, 1975), trunk crest cells formed morphologically normal presumptive cranial sensory ganglia in an environment which has no metameric mesenchymal condensations.

Many members of the neural crest population contact other tissues which might serve as substrata to promote, and perhaps direct, cell migration. The existence of attachments between the filapodia of migrating crest cells and, for example, the surface of the optic vesicle is shown in Fig. 1.14. In this preparation the embryo was transected and the eye then removed to expose the sub-epithelial surface of the crest mesenchyme. Many fine processes project from the cells which were adjacent to the optic vesicle, while those crest cells situated above the eye, beneath the superficial ectoderm, have few filapodia.

Weston (1963) has suggested that the initial pattern and oritentation of trunk neural crest cell migration is directed by contacts with the neural tube. This is based on his observation that crest cells migrate in a correct relation to the spinal cord even following 180° cord inversion and implantation either *in situ* or into the intermediate mesoderm, lateral to the somites.

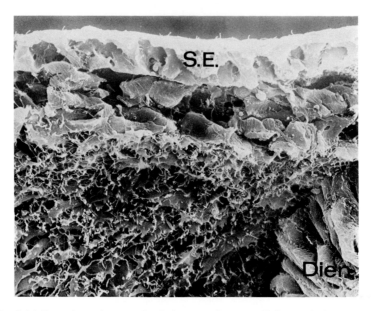

Fig. 1.14 Scanning micrograph of the neural crest cells located along the caudal surface of the optic vesicle, which has been removed. It appears that processes from these crest cells were in contact with the vesicle. (Courtesy of K. Tosney.)

The neuro-epithelium might direct the patterned movements of these crest cells by contact guidance (see Weiss, 1961), in which the arrangement of cell surface or extracellular components would influence the orientation of migrating cells. Löfberg (1976) has reported that the extracellular fibers covering the dorsal half of the axolotl spinal cord are aligned vertically, in the direction of crest cell movement. However, scanning electron microscopic studies of the chick spinal cord reveal little (Ebendal, 1977) or no (Bancroft and Bellairs, 1976; Tosney, 1978) regular orientation to the fibrous meshwork associated with this epithelium.

In vitro analyses have shown that migrating cells (Carter, 1965; Gail and Boone, 1972; Harris, 1973) and elongating nerve fibers (Letourneau, 1975) will orient towards and preferentially move upon substrata which permit contacts of greater adhesiveness. If bonds established with the spinal cord were stronger than those formed with other environmental components, migrating neural crest cells would preferentially use the neuro-epithelium as a substratum. Although some crest cells form connections with the spinal cord (Bancroft and Bellairs, 1976), and also other regions of the neuro-epithelium (Fig. 1.14), there is no evidence that the cord presents a more adhesive surface. Cells at the leading margin of the migrating crest population do not contact either the neural or the superficial epithelium, but rather penetrate the matrix in the cell-free space (Figs 1.2 and 1.3). One frequently observed fixation artifact is a separation between the crest cell mass and the brain, shown in Fig. 1.2. This indicates that the contacts among crest cells are stronger than those between crest cells and the neuro-epithelium.

These observations suggest that the spinal cord and associated extracellular material may function as a 'ground mat' (Weiss, 1961; Hay, 1968), a substratum which provides the adhesive sites necessary to support cell migration, but which does not direct the patterns of cell movement and distribution. The results of Weston's inverted-neural tube transplantations might be explained by the presence of natural or surgically created environmental conditions which promote the movement of crest cells in a plane parallel to the dorso-ventral axis of the ectopic spinal cord fragment.

Analyses of the influence of the superficial ectoderm in directing the patterned movements of cephalic crest cells have led to similar conclusions. Steffek *et al.* (in preparation) occasionally see basal lamina-associated fibers aligned in the direction of crest cell migration. However, neither the removal of pieces of the epidermis (Hamburger, 1961) nor, in amphibians, embryonic skin rotation (Hörstadius and Sellman, 1946) affect the normal migrations of crest cells.

The role of the acellular fibrous environment which emigrating crest cells enter is more difficult to analyze. Tosney (1978) was unable to detect any preferred alignment of these extracellular components, but the states that their three-dimensional organizations could have been altered during fixation.

Interactions between crest cells and matrix components are best defined in the developing cornea. Immigrating endothelial cells can use either the basal lamina of the lens or the back of the stroma as substrata for movement (Bard *et al.*, 1975), but they have been shown to adhere more tightly to the preferred substratum, the stroma (Nelson and Revel, 1975). The fibers in this matrix are initially orthogonally arranged, but the entire meshwork is highly convoluted. However, observations of the movements of endothelial cells indicate that their direction is unaffected by the orientation of the convolutions or fibers. Thus, there is no apparent basis for a contact guidance mechanism in this system, either. The only requirements for normal migration of these crest cells is, '... that the embryo create extracellular boundaries that both permit and constrain the possible routes the cells can take', (Bard *et al.*, 1975, p. 360).

Pigment cell localization is probably the most frequently analyzed aspect of neural crest cell migration and distribution. The pioneering transplantation and *in vitro* experiments of Twitty (1936, 1944, 1949) and his co-workers established that the developmental behavior of amphibian melanoblasts was the result of inter-actions both among the migrating crest cells and between them and their environ-ment. The normal migratory pathways of presumptive pigment cells can be altered if the size or integrity of the neural crest population is disrupted (Twitty and Bodenstein, 1939, 1944; Borack, 1971), indicating that the movement of these cells is not unchangeably programmed. Many experiments, most recently those of MacMillan, (1976), have indicated that the tissues with which migrating melanoblasts are associated differ greatly in their affinity for neural crest cells. Given a choice either *in vivo* or *in vitro* between dorsal hypomeric tissue, which is normally heavily pigmented in *Xenopus*, or the ventral hypomere, usually sparsely pigmented, migrating melanoblasts preferentially reside in the former.

The migration of pigment cells in avian (Rawles, 1948; Reams, 1967) and mammalian (Mayer, 1970; reviewed by Wolfe and Coleman, 1966) embryos has also been examined. Analyses of the development of the murine piebald pigment pattern (Searle, 1968; Schiable, 1969; Mayer, 1977) indicate that expression of this spotted phenotype is affected by at least 14 gene loci, some of which alter neural crest cell proliferation, maturation, or viability. Thus, while the initial pattern of distribution of melanoblasts is primarily the result of neural crest population pressure, which generally promotes dispersion, and differential affinities of local environments, other levels of control must be recognized.

1.6 CONTROL OF NEURAL CREST CYTODIFFERENTIATION

Prior to or concomitant with the expression of cytological and biochemical properties characteristic of the differentiated state, the developmental potentiality of each neural crest cell becomes restricted. This may begin prior to migration, in which case the population would be a mosaic of determined cells, or subpopulations of cells, which preferentially migrate to their appropriate terminal locations. Or, restriction of potentiality may occur during the migratory phase in response to the environmental influences described previously. Finally, neural crest cells may be multipotential until, following a period of non-specific migration, various local environments direct their differentiation. These general alternatives are not mutually exclusive, since restriction may well occur progressively over the entire course of crest cell development.

It is important that data from the observations and experiments to be discussed in the following sections be interpreted with reservation. This is because restrictive interactions, which are well defined phenomenologically, cannot at present be assayed directly. Only the subsequent expressive capacity of the cell can be analyzed, and this may be dependent upon other conditions (reviewed by Wessells, 1977). In addition, as outlined previously, in most experiments it is difficult to analyze the developmental capacity of individual neural crest *cells* rather than the potentialities of the entire population.

1.6.1 Evidence for early restriction

The experiments of Raven (1931) and Hörstadius and Sellman (1946) proved that all neural crest populations are not equipotential in amphibians. Although capable of migrating into visceral arches (Chibon, 1966), transplanted trunk crest cells are unable to form cartilage or bone. This is true also in the chick embryo (Le Douarin and Teillet, 1974).

During the initial stages of migration all neural crest cells are similar in both their gross and ultra-structural appearance (Bancroft and Bellairs, 1976), and show no indication of cytoplasmic specialization. However, Tosney (1978) reports that in the trunk some of these cells are coated with a basal lamina-like material. It is not known whether this is accumulated extracellular matrix material or secretion products of certain crest cells. Pearse and Polak (1971) find cells beside the hindbrain of the 7-somite mouse embryo that selectively sequester DOPA, which had previously been injected into the mother. Following treatment of fixed embryos with formaldehyde vapors, these cells fluoresce when exposed to ultra-violet light. Pearse and Polak assume that all these fluorescing cells are presumptive calcitonin-secreting cells. If true, this indicates the expression en route of traits characteristic of a differentiated cell. On the other hand, the large number of cells that fluoresce suggests that many, perhaps all, emigrating mouse rhombencephalic crest cells exhibit this property, in which case its significance is unclear and should be investigated further. Early migrating avian crest cells do not incorporate DOPA (Polak *et al.,* 1971).

Based on the results of neural crest extirpation experiments, several investigators have suggested that the early migrating crest cells possess developmental potentialities which differ from those cells emigrating later (Hammond, 1949; Hammond and Yntema, 1958; Andrew, 1969). However, neither transplantation studies (Weston and Butler, 1966) nor *in vitro* analyses (Cohen and Konigsberg, 1975; Maxwell, 1976), in which the properties of early and late-forming crest cells can be compared directly, have revealed any temporal differences.

Interactions between neural crest cells and their environment are necessary for the expression of certain cell phenotypes. For example, the retinal pigmented epithelium causes the crest mesenchyme to condense and chondrify adjacent to the eye, forming the sclera (Newsome, 1972; Stewart and McCallion, 1975). This interaction is facilitated by products released by the epithelium (Newsome, 1975). Similarly, differentiation of avian enteric parasympathetic neurons is dependent upon factors within the intestinal splanchnopleure (Smith *et al.,* 1977).

Both transplantation (Hörstadius and Sellman, 1946; Okada, 1955) and *in vitro* analyses (Holtfreter, 1968; Drews *et al.,* 1972; Epperlein, 1974; Corsin, 1975) have shown that chondrification of some amphibian cephalic neural crest cells requires the presence of pharyngeal endoderm. Cartilage formation always occurs in cell clusters and, while usually adjacent to the endoderm, can be affected by a diffusible agent (Holtfreter, 1968). In addition, many cells in these amphibian neural crest cultures develop into (1) pigment cells, which appear earlier and are usually scattered around the periphery of the explant, (2) neurons, which may occur singly or in cluster, or (3) fibroblasts. The presence of these cell types, with or without the presence of pharyngeal endoderm, suggests that the amphibian pre-migratory cranial neural crest is heterogeneous, with some cells committed to specific developmental pathways.

Cohen (1972) and Norr (1973) cultured avian trunk crest cells alone or in combination with other axial tissues on the chorio-allantoic membrane or *in vitro* to analyze the development of catecholamine-containing cells. They found that tissue interactions between migrating trunk crest cells and their environment, especially the somitic mesenchyme and, indirectly, the ventral neural tube, are a normal pre-requisite to sympathoblast formation. However, Cohen (1977) has recently shown that trunk crest cells maintained *in vitro* in the absence of other tissues will differ-entiate into several cell types, including sympathetic neurons with dense-core, catecholamine-containing vesicles, small intensely-fluorescent cells, neurons which possess agranular presynaptic vesicles and melanocytes. Chorio-allantoic and *in vitro* explants of cranial neural crest cells also develop both neuronal and pigment cell types (Dorris, 1941; Greenberg and Schrier, 1977).

These results have been interpreted as indicating that some premigratory crest cells are restricted to certain pathways of development and are capable of autonomous differentiation under permissive conditions. However, the possibility cannot be excluded that slight variations in the microenvironment or altered cell-to-cell relations promoted by the artificial substratum might cause multipotential cells to develop along specific pathways.

1.6.2 Evidence for pluripotentiality

The strongest evidence that some emigrating neural crest cells are uncommitted and multipotential comes from experiments in which regional differences in their develop-mental repertoire are exploited. In the chick embryo, all sympathoblasts normally originate from brachial and thoracic crest cells, parasympathetic neurons from regions rostral or caudal to these, and no neurons are formed by anterior mesencephalic and diencephalic crest cells. If the neural crest population is a mosaic of cells, each irreversibly committed to a specific fate, the developmental potentialities of the pre-migratory crest tissue from various regions would differ.

The first indications that some avian crest cells are multipotential were the results of trunk neural crest extirpation experiments (Hammond, 1949; Yntema and Hammond, 1945; Nawar, 1956). In many of these embryos crest cells adjacent to the excised region migrated into the depleted area and formed paravertebral sympatho-blasts. Similarly, catecholamine-containing cells develop in explants of cervical and cephalic neural crest cells (Chevallier, 1972; Bjerre, 1973).

The elegant transplantations of Le Douarin and her co-workers clearly prove that hindbrain crest cells will form sympathoblasts and adreno-medullary cells following transplantation in the place of the thoracic neural crest (Le Douarin and Teillet, 1974; Le Douarin, 1975). In the reciprocal grafts, thoracic crest cells contribute to enteric para-sympathetic ganglia which are morphologically and functionally normal (Le Douarin *et al.,* 1975). Moreover, pre-migratory thoracic crest cells develop acetylcholine esterase and choline acetyltransferase activity when implanted directly into the wall of the hindgut and cultured on the chorio-allantoic membrane (Smith *et al.,* 1977). Thus, in this case, the migratory experience is not an obligatory pre-requisite to cytodifferentiation. Le Douarin's transplants confirm the data of Andrew (1970) which showed that the ability to form enteric ganglia is widespread among neural crest populations.

These experiments indicate that the ability of neural crest cells to specifically form either sympathetic or parasympathetic neurons is not restricted to those regions which normally contribute to one or the other. Furthermore, the differentiation of these cells is facilitated by interactions with environmental influences encountered during migration (sympathoblasts) or at the site of terminal development (enteric neurons).

Recent analyses indicate that many neural crest cells do not become irreversibly committed to a specific autonomic function until later in development. For example, cells from dissociated neonatal mouse or rat superior cervical ganglia will, in the absence of other tissues, develop the ability to synthesize and release the catecholamine norepinephrine *in vitro* (Mains and Patterson, 1973). However, identical cells cultured in the presence of heart cells (Landis, 1976; Furshpan *et al.,* 1976; Reichardt and Patterson, 1977) or muscle cell-conditioned medium (Patterson and Chun, 1977) will synthesize and release acetylcholine. Similarly, Le Douarin *et al.* (1977) has shown that if a piece of hindgut wall, into which neural crest cells have migrated, is transplanted beside the neural tube, crest cells emigrate from the intestinal splanchnopleur into peri-axial mesenchyme and form catecholamine-containing sympathoblasts. Even cells within the 4 to 5-day avian embryonic dorsal root ganglia will, when grown *in vitro*, occasionally develop the capacity to fluoresce following treatment with formaldehyde vapors (Newgreen and Jones, 1975).

Clearly, then, the committment to differentiate as a cholinergic or catecholaminergic neuron is not made prior to the onset of migration. Le Douarin (1975) and Cohen (1977) suggest that these data indicate the existence of a subpopulation of neural crest cells restricted to a neurogenic pathway of development, but unspecified prior to migration with respect to the type of neuron they will form.

Noden (1976, 1978a,b) has examined the cytodifferentiation of heterotopically transplanted avian cephalic neural crest cells. Quail metencephalic crest cells will, following implantation in the place of chick anterior mesencephalic neural crest, form all the connective tissues normally derived from the more rostral region, as shown, for example, in Fig. 1.10. The reciprocal transplants yield identical results. In these cases the morphology of visceral arch cartilages and bones is indistinguishable from that seen in normal and control embryos. Thus, in contrast to the results of a similar set of experiments done on amphibian embryos (Hörstadius and Sellman, 1946), both the cytodifferentiation and subsequent morphogenesis of skeletal and connective tissue elements are under local environmental control.

Noden's results also indicate that the environmental components promoting the differentiation of cartilage are similar in periocular and visceral arch regions. The occasional formation of cartilage nodules by cephalic crest cells grafted into trunk regions (Le Douarin and Teillet, 1974) suggests that these factors may be common to many chondrogenic environments, although heterotopically transplanted amphibian cephalic crest cells do not respond to the peri-notochordal environment (Hörstadius and Sellman, 1946; Chibon, 1966). However, these experiments do not distinguish between exogenous chondrogenic factors, as are produced by the notochord

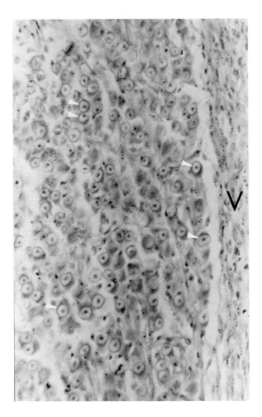

Fig. 1.15 Trigeminal ganglion from a 9-day chick embryo whose metencephalic crest cells were replaced with anterior mesencephalic crest cells excised from a quail embryo. The quail marker can be seen in sensory neurons (indicated by arrowheads), as well as in smaller cells in the ganglion and presumptive Schwann cells within the mandibular motor nerve (V).

(Holtzer and Detwiler, 1953; Kosher *et al.,* 1973; Kosher and Lash, 1975) and pigmented retinal epithelium (Newsome, 1975), and endogenous, chondroblast-produced chondrogenic agents (Cooper, 1965; Solursh and Meier, 1973).

Analysis of the ganglia formed in these chimeric avian embryos reveals that diencephalic crest cells, which normally do not form any neurons, will develop into sensory and autonomic neurons when implanted in the place of metencephalic crest cells (Noden, 1976, 1978b). The ability of heterotopically grafted crest cells to contribute to the trigeminal (Fig. 1.15) and ciliary (Fig. 1.16) ganglia, and there develop as neurons, accessory cells and Schwann sheath cells, is unequivocal.

Noden's results indicate that if, as suggested earlier, there exists a subpopulation of crest cells committed to neural pathways of development, such a subpopulation

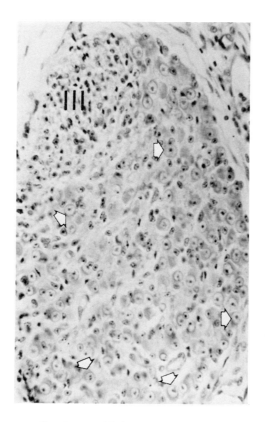

Fig. 1.16 Ciliary ganglion from a 10-day chick embryo which received a transplant similar to that in Fig. 1.15. Here too, neurons (indicated by arrows), non-neuronal cells in the ganglion and cells within the oculomotor (III) nerve contain the quail marker indicating a neural crest origin.

must be present in the anterior mesencephalic and diencephalic regions. However, these rostral neural crest cells usually do not differentiate as neurons. Thus, either this subpopulation normally fails to develop and degenerates, for which there is no evidence, or many of these neural crest cells are in fact pluripotential and capable of developing in accordance with environmental influences.

1.7 CONCLUSIONS

The experiments presented in the preceding sections indicate that the pre-migratory avian neural crest population is not a mosaic of cells which are committed to specific ontogenetic pathways. Rather, their routes of migration and subsequent cytodifferentiation are the result of interactions both among the crest cells and between them and their environment.

Exactly how the environment directs the development of each neural crest cell is not known. Weston (1971, 1972) has suggested that the specification of crest cells may be correlated with changes in crest cell-to-cell relationships which result from alterations in the environment. For example, situations in which crest cells are allowed to disperse favor expression of the pigment cell phenotype both *in vivo* and *in vitro* (Cowell and Weston, 1970). In contrast, cell aggregation, which could be promoted by a decline in the extracellular concentration of hyaluronic acid, is a pre-requisite to the differentiation of many mesenchymal derivatives. It is possible that other environmental influences present when aggregation occurs are responsible for specifying which of several potential developmental pathways are expressed. In an environment with little hyaluronic acid, which has been shown to act as an antagonist to chondrogenesis (Toole *et al.*, 1972; Solursh *et al.*, 1975), and rich in cartilage-promoting influences (chondroitin sulfate, chondromucoprotein, etc.), such as that adjacent to the pigmented retinal epithelium or in the visceral arch (Corsin, 1977), chondrogenesis would result. In the corneal milieu, stromal cells differentiate.

This integration of cytodifferentiation with early morphogenetic events may well be sufficient to account for the development of some neural crest derivatives. However, it is difficult to extrapolate from these simple interactions to the formation of heterogeneous tissues such as sensory ganglia.

Harrison (1933) warned that understanding the mechanism by which each developing cell becomes restricted to a specific fate, i.e. becomes determined, would require an appreciation of its potentialities, definition of the influences which act upon it, and characterization of the cell's response to these environmental components. In the analysis of neural crest cell development, at least the first hurdle is passed.

REFERENCES

Abercrombie, M. (1970), Contact inhibition in tissue culture. *In Vitro*, **6**, 128–142.

Abercrombie, M., Johnson, M.L. and Thomas, G.A. (1949), The influence of nerve fibers on Schwann cell migration investigated by tissue culture. *Proc. R. Soc. London, Ser. B*, **136**, 448–460.

Andrew, A. (1969), The origin of intramural ganglia. II. The trunk neural crest as a source of enteric ganglion cells. *J. Anat.*, **105**, 89–101.

Andrew, A. (1970), The origin of intramural ganglia. III. The 'vagal' source of enteric ganglion cells. *J. Anat.*, **107**, 327–336.

Andrew, A. (1971), The origin of enteric ganglia. IV. A critical review and discussion of the present state of the problem. *J. Anat.*, **108**, 169–184.

Andrew, A. (1976), An experimental investigation into the possible neural crest origin of pancreatic APUD (islet) cells. *J. Embryol. exp. Morph.*, **35**, 577–593.

Bancroft, M. and Bellairs, R. (1975), Differentiation of the neural plate and neural tube in the young chick embryo: a study by scanning and transmission electron microscopy. *Anat. Embryol.,* **147**, 309–335.

Bancroft, M. and Bellairs, R. (1976), The neural crest cells of the trunk region of the chick embryo studied by SEM and TEM. *Zoon,* **4**, 73–85.

Baid, J.B. and Hay, E.D. (1975), The behavior of fibroblasts from the developing avian cornea. *J. Cell Biol.,* **67**, 400–418.

Bard, J.B.L., Hay, E.D. and Meller, S.M. (1975), Formation of the endothelium of the avian cornea: a study of cell movement *in vivo. Dev. Biol.,* **42**, 334–361.

Billings-Gagliardi, S., de F. Webster, H. and O'Connell, M.F. (1974), *In vivo* and electron microscopic observation of Schwann calls in developing tadpole nerve fibers. *Am. J. Anat.,* **141**, 375–391.

Bjerre, B. (1973), The production of catecholamine-containing cells *in vitro* by young chick embryos studied by the histochemical fluorescence method. *J. Anat.,* **115**, 119–131.

Borack, L. (1971), Gene action on proliferation and migration in the developing neural crest of black and white axolotls, *Ambystoma mexicanum, Shaw. J. exp. Zool.,* **179**, 289–298.

van Campenhout, E. (1937), Le rôle de la crête ganglionnaire dans la formation de mésenchyme céphalique chez l'embryon de poulet. *C. R. Soc. Biol.,* **124**, 1005–1006.

Carter, S.B. (1965), Principles of cell motility: the direction of cell movement and cancer invasion. *Nature,* **108**, 1183–1187.

Chamley, J.H. and Dowel, J.J. (1975), Specificity of nerve fiber attraction to autonomic effector organs in tissue culture. *Exp. Cell Res.,* **90**, 1–7.

Chamley, J.H., Galler, I. and Burnstock, G. (1973), Selective growth of sympathetic nerve fibers to explants of normally densely innervated autonomic effector organs in tissues culture. *Dev. Biol.,* **31**, 362–379.

Chevallier, A. (1972), Localisation et durée des potentialités médullo-surénaliennes des crêtes neurales chez le poulet. *J. Embryol. exp. Morph.,* **27**, 603–614.

Chibon, P. (1966), Analyse expérimentale de la régionalisation et des capacités morphogénetiques de la crête neurale chez l'amphibien urodèle *Pleurodeles waltlii, Michah. Mem. Soc. Zool., France,* **36**, 4–117.

Church, R.L., Tanzer, M.L. and Pfeiffer, S.E. (1973), Collagen and procollagen production by a clonal line of Schwann cells. *Proc. natn. Acad. Sci., U.S.A.,* **70**, 1943–1946.

Cohen, A.M. (1972), Factors directing the expression of sympathetic nerve traits in cells of neural crest origin. *J. exp. Zool.,* **179**, 167–182.

Cohen, A.M. (1977), Independent expression of the adrenergic phenotype by neural crest *in vitro. Proc. natn. Acad. Sci., U.S.A.,* **74**, 2899–2903.

Cohen, A.M. and Hay, E.D. (1971), Secretion of collagen by embryonic neuro-epithelium at the time of spinal cord – somite interaction. *Dev. Biol.,* **26**, 578–605.

Cohen, A.M. and Konigsberg, A.M. (1975), A clonal approach to the problem of neural crest determination. *Dev. Biol.,* **46**, 262–280.

Conrad, G.W. (1970), Collagen and mucopolysaccharide biosynthesis in the developing chick cornea. *Dev. Biol.,* **21**, 292–317.

Conrad, G.W., Hart, G.W. and Chen, Y. (1977), Differences *in vitro* between fibro-
blast-like cells from the cornea, heart, and skin of embryonic chicks.
J. Cell Sci., **26**, 119–137.

Cooper, G.W. (1965), Induction of somitic chondrogenesis by cartilage and notochord:
a correlation between inductive activity and specific stages of cytodifferentiation.
Dev. Biol., **12**, 185–212.

Corsin, J. (1975), Différenciation *in vitro* de cartilage à partir des crêtes neurales
céphaliques chez *Pleurodeles waltlii, Michah. J. Embryol. exp. Morph.,* **33**,
335–342.

Corsin, J. (1977), Le matériel extracellulaire au cours du dévelopment du chondro-
crâne des amphibiens: mise en place et constitution. *J. Embryol. exp. Morph.,*
38, 139–149.

Coughlin, M.D. (1975), Target organ stimulation of parasympathetic nerve growth
in the developing mouse submandibular gland. *Dev. Biol.,* **43**, 140–158.

Coulombre, A.J. (1965), Problems in corneal morphogenesis. *Advances in Morphogenesis,* **4**, 81–109.

Cowell, L.A. and Weston, J.A. (1970), An analysis of melanogenesis in cultured chick
embryo spinal ganglia. *Dev. Biol.,* **22**, 670–697.

Dahl, I.M., Johnsen, W., Anseth, A. and Prydz, H. (1974), The synthesis of glycosaminoglycans by corneal stroma cells *in vitro. Exp. Cell Res.,* **88**, 193–197.

Davis, E.M. (1977), Movement of neural crest cells under different culture conditions.
J. Cell Biol., **75**, 158a.

Derby, M.A., Pintar, J.E. and Weston, J.A. (1976), Glycosaminoglycans and the
development of the trunk neural crest. *J. gen. physiol.,* **68**, 4a.

Detwiler, S.R. (1934), An experimental study of spinal nerve segmentation in
Amblystoma with reference to the plurisegmental contributions to the
brachial plexus. *J. exp. Zool.,* **67**, 395–441.

Detwiler, S.R. (1937), Observations upon the migration of neural crest cells, and
upon the development of the spinal ganglia and vertebral arches in *Ambystoma.*
Am. J. Anat., **61**, 63–94.

di Virgilio, G., Lavenda, N. and Worden, J.L. (1967), Sequences of events in neural
tube closure and the formation of neural crest in the chick embryo. *Acta Anat.,*
68, 127–146.

Dodson, T.W. and Hay, E.D. (1971), Secretion of collagenous stroma by isolated
epithelium grown *in vitro. Exp. Cell Res.,* **65**, 215–220.

Dorris, F. (1941), The behavior of chick neural crest in grafts to the chorio-allantoic
membrane. *J. exp. Zool.,* **86**, 205–233.

Drews, U., Kocher-Becker, U. and Drews, U. (1972), Die Induktion von Kiemenknorpel
aus Lopfneuralleistenmaterial durch präsumptive Kiemendarm in der Gewebekultur
und das Bewegungsverhalten der Zellen. *Wilhelm Roux' Arch. Entwicklungsmech.
Org.,* **171**, 17–37.

Ebendal, T. (1977), Extracellular matrix fibrils and cell contacts in the chick embryo.
Cell Tiss. Res., **175**, 439–458.

Epperlein, H.H. (1974), The ectomesenchymal-endodermal interaction-system
(EEIS) of *Triturus alpestris* in tissue culture. I. Observations on attachment,
migration, and differentiation of neural crest cells. *Differentiation,***2**,
151–168.

Fisher, M. and Solursh, M. (1977), Glycosaminoglycan localization and role in maintenance of tissue spaces in the early chick embryo. *J. Embryol. exp. Morph.,* **42**, 195–207.

Fontaine, J., Le Lievre, C. and N.M. Le Douarin (1977), What is the developmental fate of the neural crest cells which migrate into the pancreas in the avian embryo? *Gen. comp. Endocrin.,* **33**, 394–404.

Furschpan, E.J., MacLeish, P.R., O'Lague, P.H. and Potter, D.D. (1976), Chemical transmission between rat sympathetic neurons and cardiac myocytes developing in microcultures: evidence for cholinergic, adrenergic, and dual-function neurons. *Proc. natn. Acad. Sci., U.S.A.,* **73**, 4225–4229.

Gail, M.H. and Boone, C.W. (1972), Cell-substrate adhesivity: a determinatant of cell motility. *Exp. Cell Res.,* **70**, 33–40.

Greenberg, J.H. and Pratt, R.M. (1977), Glycosaminoglycan and glycoprotein synthesis by cranial neural crest cells *in vitro. Cell Diff.,* **6**, 119–132.

Greenberg, J.H. and Schrier, B.K. (1977), Development of choline acetyltransferase activity in chick cranial neural crest cells in culture. *Dev. Biol.,* **61**, 86–93.

Hamburger, V. (1961), Experimental analysis of the dual origin of the trigeminal ganglion in the chick embryo. *J. exp. Zool.,* **148**, 91–124.

Hammond, W.S. (1949), Formation of the sympathetic nervous system in the chick embryo following removal of the thoracic neural tube. *J. comp. Neurol.,* **91, 67–86.**

Hammond, W.S. and Yntema, C.L. (1947), Depletions of the thoraco-lumbar sympathetic system following removal of neural crest in the chick. *J. comp. Neurol.,* **86**, 237–266.

Hammond, W.S. and Yntema, C.L. (1958), Origin of the ciliary ganglion in the chick. *J. comp. Neurol.,* **110**, 367–390.

Hammond, W.S. and Yntema, C.L. (1964), Depletions of pharyngeal arch cartilages following extirpation of cranial neural crest in chick embryos. *Acta Anat.,* **56**, 21–34.

Harris, A. (1973), Behavior of cultured cells on substrata of variable adhesiveness. *Exp. Cell Res.,* **77**, 285–297.

Harrison, R.G. (1906), Further experiments on the development of peripheral nerves. *Am. J. Anat.,* **5**, 121–131.

Harrison, R.G. (1924), Neuroblast versus sheath cell in the development of peripheral nerves. *J. comp. Neurol.,* **37**, 123–205.

Harrison, R.G. (1933), Some difficulties of the determination problem. *Am. Nat.,* **67**, 306–321.

Hay, E.D. (1968), Organization and fine structure of epithelium and mesenchyme in the developing chick embryo. In: *Epithelial-Mesenchymal Interactions* (Fleischmajer, R. and Billingham, R.E., eds) The Williams and Wilkins Co., Baltimore, pp. 31–55.

Hay, E.D. and Meier, S. (1974), Glycosaminoglycan synthesis by embryonic inductors: neural tube, notochord, and lens. *J. Cell Biol.,* **62**, 889–898.

Hay, E.D. and Revel, J.P. (1969), Fine Structure of the Developing Avian Cornea. *Monographs in Developmental Biology,* Vol. 1. (Wolsky, A. and Chen, P.S., eds) S. Karger AG, Basel.

Heaysman, J.E.M. and Pegrum, S.M. (1973), Early contacts between fibroblasts: an ultrastructural study. *Exp. Cell Res.,* **78**, 71–78.

Hillman, N.H. and Hillman, R. (1965), Chick cephalogenesis. The normal development of the cephalic region of stages 3 through 11 chick embryos. *J. Morph.,* **116**, 357–370.

Holmdahl, D.E. (1928), Die Entstehung und weitere Entwicklung der Neuralleiste (Ganglienleiste) bei Vögelen und Saugtieren. *Z. Mikr. Anat. Forsch.,* **14**, 99–298.

Holtfreter, J. (1968), On mesenchyme and epithelia in inductive and morphogenetic processes. In: *Epithelial-Mesenchymal Interactions* (Fleischmajer, R. and Billingham, R.E., eds) The Williams and Wilkins Co., Baltimore, pp. 1–30.

Holtzer, H. and Detwiler, S.R. (1953), An experimental analysis of the development of the spinal column. III. Induction of skeletogenous cells. *J. exp. Zool.,* **123**, 335–366.

Hörstadius, S. (1950), *The Neural Crest.* Oxford University Press, London.

Hörstadius, S. and Sellman, S. (1946), Experimentelle Unterschungen über die Determination des Knorpeligen Kopfskelettes bei Ubrodelen. *Nova Acta Soc. Scient. Upsaliensis, Ser.* 4, **13**, 1–170.

Johnston, M.C. (1966), A radioautographic study of the migration and fate of cranial neural crest cells in the chick embryo. *Anat. Rec.,* **156**, 143–156.

Johnston, M.C. (1974), Regional embryology: aspects relevant to the embryogenesis of craniofacial malformations. In: *Proceedings of the International Conference of Craniofacial Malformations* (Converse, J.M. and Pruzansky, S., eds) C.V. Mosby, St. Louis.

Jonston, M.C. and Hazelton, R.D. (1972), Embryonic origins of facial structures related to oral sensory and motor function. In: *Third Symposium on Oral Sensation and Perception: The Mouth of the Infant.* (Bosma, J.B., ed.) Chas. C. Thomas, Springfield.

Johnston, M.C. and Listgarten, M.A. (1972), Observations on the migration, interaction, and early differentiation of orofacial tissues. In: *Developmental Aspects of Oral Biology* (Slavkin, H.C. and Bavetta, L.A., eds) Academic Press, New York.

Kosher, R.A. and Lash, J.W. (1975), Notochordal stimulation of *in vitro* somite chondrogenesis before and after enzymatic removal of perinotochordal materials. *Dev. Biol.,* **42**, 362–378.

Kosher, R.A., Lash, J.W. and Minor, R.W. (1973), Environmental enhancement of *in vitro* chondrogenesis. IV. Stimulation of somite chondrogenesis by exogenous chondromucoprotein. *Dev. Biol.,* **35**, 210–220.

Landis, S.C. (1976), Rat sympathetic neurons and cardiac myocytes developing in microcultures: correlation of the fine structure of endings with neurotransmitter function in single neurons. *Proc. natn. Acad. Sci., U.S.A.,* **73**, 4220–4224.

Le Douarin, N.M. (1969), Particularités du noyau interphasique chez la caille Japonaise (*Coturnix coturnix japonica*). *Bull. Biol. Fr. Belg.,***103**, 435–442.

Le Douarin, N.M. (1971), Charactéristiques ultrastructurales du noyau intephasique chez la caille et chez le poulet et utilisation de cellules de caille comme 'marquers biologiques' en embryologie expérimentale. *Ann. Embryol. Morph,,* **4**, 125–135.

Le Douarin, N.M. (1975), Extracellular factors controlling the migration and differentiation of the ganglioblasts of the autonomic nervous system. In: *Extracellular Matrix Influences on Gene Expression* (Slavkin, H.C. and Greulich, R.C., eds) Academic Press, New York, pp. 591–600.

Le Douarin, N.M. and Le Lièvre, C. (1970), Démonstration de l'origine neurale des cellules à calcitonine du corps ultimobranchial chez l'embryon de poulet. *C. R. Acad. Sci.,* Paris, **270**, 2857–2860.

Le Douarin, N.M. and Teillet, M.N. (1970), Sur quelques aspects de la migration des cellules neurales chez l'embryon de poulet étudiée par la méthode des greffes hétérospecifique de tube nerveux. *C. R. Seances Soc. Biol. Filiales,* **164**, 390–397.

Le Douarin, N.M. and Teillet, M.A. (1971), Localisation, par la méthode greffes interspécifiques du territoire neural dont dérivent les cellules adrénales surrenaliennes chez l'embryon d'oiseau. *C. R. Acad. Sci.,* Paris, **272**, 481–484.

Le Douarin, N.M. and Teillet, M.A. (1973), The migration of neural crest cells to the wall of the digestive tract in avian embryos. *J. Embryol. exp. Morph.,* **30**, 31–48.

Le Douarin, N.M. and Teillet, M.A. (1974), Experimental analysis of the migration and differentiation of neuroblasts of the autonomic nervous system and of neurectodermal mesenchymal derivatives, using a biological cell marking technique. *Dev. Biol.,* **41**, 162–184.

Le Douarin, N.M.,Renaud, D., Teillet, M.A. and Le Douarin, G. (1975), Cholinergic differentiation of presumptive adrenergic neuroblasts in interspecific chimeras after heterotopic transplantations. *Proc. natn. Acad. Sci., U.S.A.,* **72**, 728–732.

Le Douarin, N.M., Teillet, M.A. and Le Lièvre, C. (1977), Influence of the tissue environment on the differentiation of neural crest cells. In: *Cell and Tissue Interactions.* (Lash, J.W. and Burger, M., eds) Raven Press, New York.

Lehman, F. (1927), Further studies on the morphogenetic role of the somites in the development of the nervous system of amphibians. The arrangement and differentiation of spinal ganglia in *Pleurodeles waltlii. J. exp. Zool.,* **49**, 93–142.

Le Lievre, C. (1974), Rôle des cellules mesectodermiques issues des crêtes neurales céphaliques dans la formation des arcs branchiaux et du squelette viscéral. *J. Embryol. exp. Morph.,* **31**, 453–577.

Le Lièvre, C. and Le Douarin, N.M. (1975), Mesenchymal derivatives of the neural crest: analysis of chimeric quail and chick embryos. *J. Embryol. exp. Morph.,* **34**, 125–154.

Letourneau, P.C. (1975), Cell-to-substratum adhesion and guidance of axonal elongation. *Dev. Biol.,* **44**, 92–101.

Löfberg, J. (1976), Scanning and transmission electron microscopy of early neural crest migration and extracellular fiber systems of the amphibian embryo. *J. Ultrastruct. Res.,* **54**, 484a.

Low, F.N. (1970), Interstitial bodies in the early chick embryo. *Am. J. Anat.,* **128**, 45–56.

MacMillan, G. (1976), Melanoblast — tissue interactions and the development of pigment pattern. *J. Embryol. exp. Morph.,* **35**, 463–484.

Mains, R.E. and Patterson, P.H. (1973), Primary cultures of dissociated sympathetic neurons. I. Establishment of long-term growth in culture and studies of differentiated properties. *J. Cell Biol.,* **59,** 329–345.

Manasek, F.J. (1975), The extracellular matrix: a synamic component of the developing embryo. In: *Current Topics in Developmental Biology,* Vol. 10 (Moscona, A. and Montroy, A., eds) Academic Press, New York, pp. 35–102.

Manasek, F.J. (1976), Heart development: interactions involved in cardiac morphogenesis. In: *The Cell Surface in Animal Embryogenesis and Development.* (Poste, G. and Nicholson, G.L., eds) Elsevier/North Holland Biomedical Press, pp. 545–598.

Manasek, F.J. and Cohen, A.M. (1977), Anionic glycopeptides and glycosaminoglycans synthesized by embryonic neural tube and neural crest. *Proc. natn. Acad. Sci., U.S.A.,* **74,** 1057–1061.

Maxwell, G.D. (1976), Cell cycle changes during neural crest cell differentiation *in vitro. Dev. Biol.,* **49,** 66–79.

Mayer, T.C. (1970), A comparison of pigment cell development in albino, steel, and dominant spotting mutant mouse embryos. *Dev. Biol.,* **23,** 297–309.

Mayer, T.C. (1977), Enhancement of melanocyte development from piebald neural crest by a favorable tissue environment. *Dev. Biol.,* **56,** 255–262.

Meade, P.A. and Norr, S.C. (1977), The onset of neural crest cell migration: The basal laminae. *J. Cell Biol.,* **75,** 48a.

Meier, S. and Hay, E.D. (1974), Control of corneal differentiation by extracellular materials. Collagen as a promoter and stabilizer of epithelial stroma production. *Dev. Biol.,* **38,** 249–270.

Nawar, G. (1956), Experimental analysis of the origin of the autonomic ganglia in the chick embryo. *Am. J. Anat.,* **99,** 473–506.

Nelson, G.A. and Revel, J.P. (1975), Scanning electron microscopic study of cell movements in the corneal endothelium of the avian embryo. *Dev. Biol.,* **42,** 315–333.

Newgreen, D.F. and Jones, R.O. (1975), Differentiation *in vitro* of sympathetic cells from chick embryo sensory ganglia. *J. Embryol. exp. Morph.,* **33,** 43–56.

Newsome, D.A. (1972), Cartilage induction by retinal pigmented epithelium of the chick embryo. *Dev. Biol.,* **27,** 575–579.

Newsome, D.A. (1975), *In vitro* induction of cartilage in embryonic chick neural crest cells by products of retinal pigmented epithelium. *Dev. Biol.,* **49,** 496–507.

Noden, D.M. (1973), The migratory behavior of neural crest cells. In: *Fourth Symposium on Oral Sensation and Perception: Development in the Fetus and Infant.* (Bosma, J., ed) U.S. Dept. H.E.W., Bethesda.

Noden, D.M. (1975), An analysis of the migratory behavior of avian cephalic neural crest cells. *Dev. Biol.,* **42,** 106–130.

Noden, D.M. (1976), Cytodifferentiation in heterotropically transplanted neural crest cells. *J. gen. Physiol.,* **68,** 13a.

Noden, D.M. (1978a), The control of avian cephalic neural crest cytodifferentiation. I. Skeletal and connective tissues. (submitted).

Noden, D.M. (1978b), The control of avian cephalic neural crest cytodifferentiation. II. Neural tissues. (submitted).

Norr, S.C. (1973), *In vitro* analysis of sympathetic neuron differentiation from chick neural crest cells. *Dev. Biol.,* **34**, 16–38.

Okada, E.I. (1955), Isolationsversuche zur Analyse der Knorpelbildung aus Neuralleistenzellen bei Urodelenkeim. *Mem. Coll. Sci., Kyoto, Ser. B,* **22**, 23–28.

Patterson, P.H. and Chun, L.L.Y. (1977), The induction of acetylcholine synthesis in primary cultures of dissociated rat sympathetic neurons. II. Developmental aspects. *Dev. Biol.,* **60**, 473–481.

Pearse, A.G.E. (1969), The cytochemistry and ultrastructure of polypeptide hormone-producing cells of the APUD series and the embryologic, physiologic and pathologic implications of the concept. *J. Histochem. Cytochem.,* **17**, 303–313.

Pearse, A.G.E. and Polak, J.M. (1971), Cytochemical evidence for the neural crest origin of mammalian ultimobranchial C cells. *Histochemie,* **27**, 96–102.

Pearse, A.G.E., Polak, J.M., Rost, F.W.D., Fontaine, J., Le Lièvre, C. and Le Douarin, N.M. (1973), Demonstration of the neural crest origin to type I (APUD) cells in the avian carotid body using a cytochemical marker system. *Histochemie,* **34**, 191–203.

Pictet, R.L., Rall, L.B., Phelps, P. and Rutterm, W.J. (1976), The neural crest and the origin of the insulin-producing and other gastro-intestinal hormone-producing cells. *Science,* **101**, 191–192.

Polak, J.M., Rost, F.W.D. and Pearse, A.G.E. (1971), Fluorogenic amine tracing of neural crest derivatives forming the adrenal medulla. *Gen. Comp. Endocrinol.,* **16**, 132–136.

Pollack, E.D. and Liebig, V. (1977), Differentiating limb tissue affects neurite outgrowth in spinal cord cultures. *Science,* **197**, 899–900.

Pratt, R.M., Larsen, M.A. and Johnston, M.C. (1975), Migration of cranial neural crest cells in a cell-free hyaluronate-rich matrix. *Dev. Biol.,* **44**, 298–305.

Pratt, R.M., Morriss, G. and Johnston, M.C. (1976), The source, distribution, and possible role of hyaluronate in the migration of chick cranial neural crest cells. *J. gen. Physiol.,* **68**, 15a.

Raven, C.P. (1931), Zur Entwicklung der Ganglienleiste. I. Kinematic der ganglienleisten Entwicklung bei der Urodelen. *Wilhelm Rous' Arch. Entwicklungsmech. Org.,* **125**, 210–292.

Rawles, M.E. (1948), Origin of melanohores and their role in development of color patterns in vertebrates. *Physiol. Rev.,* **28**, 383–408.

Reams, W.M., Jr. (1967), Pigment cell population pressure within the skin and its role in the pigment cell invasion of extra-epidermal tissues. *Adv. Biol. Skin,* **8**, 489–501.

Reichardt, L.F. and Patterson, P.H. (1977), Neurotransmitter synthesis and uptake by isolated sympathetic neurons in microcultures. *Nature,* **270**, 147–150.

Schiable, R.H. (1969), Clonal distribution of melanocytes in piebald-spotted and variegated mice. *J. exp. Zool.,* **172**, 181–200.

Searle, A.G. (1968), *Comparative Genetics of Coat Color in Mammals.* Logos/ Academic, New York.

Sidman, R.L. (1974), Contact interactions among developing mammalian brain cells. In: *The Cell Surface in Development.* (Moscona, A., ed.) John Wiley & Sons, New York, pp. 221–254.

Smith, J., Cochard, P. and Le Douarin, N.M. (1977), Development of choline acetyltransferase and cholinesterase activities in enteric ganglia derived from presumptive adrenergic and cholinergic levels of the neural crest. *Cell. Diff.,* **6**, 199–216.

Solursh, M. (1976), Glycosaminoglycan synthesis in the chick gastrula. *Dev. Biol.,* **50**, 525–530.

Solursh, M. and Meier, S. (1973), A conditioned medium (CM) factor produced by chondrocytes that promotes their own differnetiation. *Dev. Biol.,* **30**, 279–289.

Solursh, M., Vaerewyck, S.A. and Reiter, R.S. (1975), Depression by hyaluronic acid of glycosaminoglycan synthesis by cultured chick embryo chondrocytes. *Dev. Biol.,* **41**, 233–244.

Speidel, C.C. (1932), Studies of living nerves. I. The movements of individual sheath cells and nerve sprouts correlated with the process of myelin sheath formation in amphibian larvae. *J. exp. Zool.,* **61**, 279–331.

Spemann, H. (1938), *Embryonic Development and Induction.* Yale University Press, New Haven.

Stewart, P.A. and McCallion, D.J. (1975), Establishment of the scleral cartilage in the chick. *Dev. Biol.,* **46**, 383–389.

Teillet, M.A. (1971), Recherches sur le mode de migration et la différenciation des mélanoblastes cutanés chez l'embryon d'oiseau. *A. Embryol. Morph.,* **4**, 95–109.

Toole, B.P. (1972), Hyaluronate turnover during chondrogenesis in the developing chick limb and axial skeleton. *Dev. Biol.,* **29**, 321–329.

Toole, B.P. (1976), Morphogenetic role of glycosaminoglycans (acid mucopolysaccharides) in brain and other tissues. In: *Neuronal Recognition.* (Barondes, S.H., ed.) Plenum Press, New York, pp. 275–329.

Toole, B.P. and Trelstad, R.L. (1971), Hyaluronate production and removal during corneal development in the chick. *Dev. Biol.,* **26**, 28–35.

Toole, B.P., Jackson, G. and Gross, J. (1972), Hyaluronate in morphogenesis: inhibition of chondrogenesis *in vitro. Proc. natn. Acad. Sci., U.S.A.,* **69**, 1384–1386.

Tosney, K.T. (1978), The early migration of neural crest cells in the trunk region of the avian embryo: an SEM–TEM study. *Dev. Biol.,* in press.

Trelstad, R.L. (1977), Mesenchymal cell polarity and morphogenesis of chick cartilage. *Dev. Biol.,* **59**, 153–163.

Trelstad, R.L. and Coulombre, A.J. (1971), Morphogenesis of the collagenous stroma in the chick cornea. *J. Cell Biol.,* **50**, 849–858.

Trelstad, R.L., Hay, E.D. and Revel, J.P. (1967), Cell contact during early morphogenesis in the chick embryol Dev. Biol., **16**, 78–106.

Trinkaus, J.P. (1969), *Cells Into Organs.* Prentiss-Hall, Engelwood Cliffs, N.J.

Trinkaus, J.P. (1976), On the mechanisms of metazoan cell movements. In: *The Cell Surface in Animal Embryogenesis and Development.* (Post, G. and Nicholson, G.L., eds) Elsevier/North Holland Biomedical Press. pp. 225–329.

Triplett, E.L. (1958), The development of the sympathetic ganglia, sheath cells, and meninges in Amphibians. *J. exp. Zool.,* **138**, 283–311.

Twitty, V.C. (1936), Correlated genetic and embryological experiments of *Triturus.* I. Hybridization, and II. Transplantation. *J. exp. Zool.,* **74**, 239–302.

Twitty, V.C. (1944), Chromatophore migration as a response to mutual influences of the developing pigmented cells. *J. exp. Zool.,* **95**, 259–290.

Twitty, V.C. (1949), Developmental analysis of amphibian pigmentation. *Growth,* **13**, (Suppl. 9) 133–161.

Twitty, V.C. and Bodenstein, D. (1939), Correlated genetic and embryological experiments on *Triturus. J. exp. Zool.,* **81**, 357–398.

Twitty, V.C. and Bodenstein, D. (1944), The effect of temporal and regional differentials on the development of grafted chromatophores. *J. exp. Zool.,* **95**, 213–231.

Twitty, V.C. and Niu, M.C. (1948), Causal analysis of chromatophore migration. *J. exp. Zool.,* **108**, 405–437.

Twitty, V.C. and Niu, M.C. (1954), The motivation of cell migration studied by isolation of embryonic pigment cells singly and in small groups *in vitro. J. exp. Zool.,* **125**, 541–574.

Weiss, P.A. (1961), Guiding principles in cell locomotion and aggregation. *Exp. Cell Res.,* Suppl. 8, 260–281.

Wessells, N.K. (1977), *Tissue Interactions and Development.* W.A. Benjamin, Inc., Menlo Park, Calif.

Weston, J.A. (1963), A radioautographic analysis of the migration and localization of trunk neural crest cells in the chick. *Dev. Biol.,* **6**, 279–310.

Weston, J.A. (1970), The migration and differentiation of neural crest cells. *Adv. Morphogenesis,* **8**, 41–114.

Weston, J.A. (1971), Neural crest cell migration and differentiation. In: *Cellular Aspect of Growth and Differentiation in Nervous Tissue* (Pease, D., ed.) U.C.L.A. Forum in Medical Sciences, **14**, 1–19.

Weston, J.A. (1972), Cell interaction in neural crest development. In: *Cell Interactions* (Silvestri, L.G., ed.) American Elsevier Publishing Co., Inc., New York, pp. 286–292.

Weston, J.A. and Butler, S.L. (1966), Temporal factors affecting localization of neural crest cells in the chicken embryo. *Dev. Biol.,* **14**, 246–266.

Wolfe, H.G. and Coleman, D.L. (1966), Pigmentation. In: *Biology of the Laboratory Mouse* (Green, E.L., ed.) McGraw-Hill Book Co., New York, pp. 405–425.

Wood, P.M. and Bunge, R.R. (1975), Evidence that sensory axons are mitogenic for Schwann cells. *Nature,* **256**, 662–664.

Yntema, C.L. (1944), Experiments on the origin of the sensory ganglia of the facial nerve in the chick. *J. comp. Neurol.,* **81**, 147–167.

Yntema, C.L. and Hammond, W.S. (1945), Depletions and abnormalities in the cervical sympathetic system of the chick following extirpation of neural crest. *J. exp. Zool.,* **100**, 237–263.

Yntema, C.L. and Hammond, W.S. (1954), The origin of intrinsic ganglia of trunk viscera from vagal neural crest in the chick. *J. comp. Neurol.,* **101**, 515–542.

2 The Problem of Specificity in the Formation of Nerve Connections

R.M. G A Z E

Specificity of Embryological Interactions
(*Receptors and Recognition,* Series B, Volume 4)
Edited by D.R. Garrod
Published in 1978 by Chapman and Hall, 11 New Fetter Lane, London EC4P 4EE
© Chapman and Hall

2.1 THE NATURE OF THE PROBLEM

For thirty years or so the word 'specificity' has been much used but under-defined in relation to the developing nervous system, and this has led to considerable confusion. Thus we may talk of specificity of nerve connections and of neuronal specificity; and these two usages have quite different meanings. By 'specificity of nerve connections' we mean that nerve fibres make predictably particular and selective connections. The phrase, 'neuronal specificity', however, has other connotations. It is used to denote the hypothesis, put forward by Sperry (1943; 1944; 1945; 1951; 1963; 1965) to account for the formation of specific or particular nerve connections. I intend to discuss this hypothesis, and consider various of its ramifications, later; here it is merely necessary to point out that, whereas the mechanism proposed in the hypothesis of neuronal specificity is one way to establish specific nerve connections, it is by no means the only way. Specific nerve connections could well be established by a variety of other mechanisms, some of which will also be discussed below. In this chapter I will outline the possibilities as I see them at present, and attempt to assess these in terms of probability.

The numbers of neurons in a nervous system can be very large, both in vertebrates and in invertebrates. A figure of 10^{12} is sometimes quoted for the brain of man and the figure would be smaller but still very large for the frog. Each neuron can receive input connections from several hundred or several thousand fibres from other neurons and all the evidence available at present indicates that the greater part (or all) of these interconnections are formed according to plan. As far as can be demonstrated by available techniques, they show order. In the present context this may mean, firstly, that a particular distribution of connections is such that we recognize it to be non-random. Secondly, that there is a distribution of connections that recurs recognizably from one animal to another (an example could be the map of retinal fibre connections on the optic tectum). A patterned or non-random distribution need not, however, be 'correct', in that it may not be what we would expect to find in a normal animal. It is possible to generate, by surgical means, a connection pattern that is ordered but incorrect. In such a case, while not necessarily repeatable from one animal to another the order may show itself in preservation of the polarity of the spatial distribution of connections, and in maintenance of the internal arrangement of the fibres within the boundaries of the array of pre-synaptic fibres.

Patterns of ordered nerve connections are found throughout the central nervous system. The most extensively investigated patterns are probably those found in the visual, auditory and somatosensory systems, as well as those in the cerebellum and the hippocampus. One important peculiarity of many of these systems of connections is that, not only are they ordered in the sense previously defined, they are also

53

continuously ordered and form mappings of the pre-synaptic array of fibres on to the postsynaptic array of cells or dendrites. In a continuously ordered mapping (Gaze and Hope, 1976) neurons which are neighbours in one array project to, or receive from, neurons that are neighbours in the other array. Thus in such systems of inter-connecting neurons, or mappings, of which the retinotectal system in lower vertebrates is a good example, neighbourhood relationships are preserved; and this fact, as we will see later, may well be vitally important when we come to consider in detail some of the causal mechanisms that have been proposed.

We are faced, therefore, with the existence of very many groups of ordered connections within the adult or mature nervous system, including many continuously ordered mappings; and in the young embryo there are no neural connections at all, because there are no nerve cells. Concurrently with development of the embryo through the neurula stage, neurons begin to appear and also neural inter-connections. The problem we are concerned with is: What factors and mechanisms are responsible for the development of the orderly connection patterns we later see? How are neuronal mappings set up?

An ordered neuronal projection is not the same thing as ordered neural connections. The projection is a distribution of nerve fibres; these will normally make synaptic connections, in which case the two terms becomes synonymous. However, in principle, the ordered projections of fibres need not necessarily make any synaptic connections; and in such a case we have to distinguish and separate the factors leading up to an ordered projection from those giving connections. A possible example of this situation is the development of the visual system in the chick, where optic fibres appear to give rise to an ordered projection across the tectum before any retionotectal synapses are formed (Crossland *et al.*, 1975). Such a situation may be contrasted with that in the developing amphibian, where optic fibres form synapses as they arrive at the growing tectum, over a period of six weeks or more. As I shall argue in a later section, the establishment of ordered connections may be dependent on the initial existence of an ordered fibre projection. Since, in most cases, ordered projections lead rapidly to ordered connections, I shall generally use the terms as equivalent to one another.

We can classify the various possible mechanisms for the establishment of ordered mappings as follows (Gaze and Hope, 1976): Target-affinity mechanisms, time-position mechanisms, fibre-sorting mechanisms and other mechanisms.

2.2 TARGET-AFFINITY MECHANISMS

As the name indicates, with target-affinity mechanisms we imply that the nerve connections that form during development do so as a result of differential affinities between neurons. When we consider the formation of ordered connection-maps between arrays of neurons (for instance the retina and the optic tectum), the hypothesis is that there are differences in the affinity of one fibre in the pre-synaptic array for cells in the postsynaptic array, and vice-versa. These differences then account for the formation of the map between the arrays.

At the present time the most widely held hypotheis concerning the formation of nerve connections is an affinity mechanism: the chemo-affinity hypothesis of Sperry, frequently called the hypothesis of neuronal specificity. The workings of such systems may appear very simple in principle; since each pre- and postsynaptic element bears its characteristic chemo-affinity label, the setting up of connections presents no problem. One merely has to ensure that the incoming fibres get to the right general part of the nervous system and the correct connections will be formed automatically. However, the affinity mechanism may make very heavy demands on genetic information; each cell has to be labelled and there are a lot of cells.

From his studies on regnerating optic nerves in urodele and anuran juveniles and adults, Sperry (1943; 1944; 1945; 1951; 1963; 1965) was led to propose that:

(1) In addition to their obvious morphological differentiation, the developing ganglion cells in the retina undergo a form of chemical cytodifferentiation, related to their position in the cell sheet, such that eventually each ganglion cell comes to possess a unique cytochemical specificity, or label. Thus each mature ganglion cell is distinct from all others and the early acquisition of these differences is under genetic control.
(2) A comparable series of developmental events occurs in the growing tectum resulting in each tectal neuron (destined to receive input from an optic fibre) acquiring a place-related cytochemical specificity. These tectal specificities, also under genetic control, match those in the retina, so that;
(3) When the retinal ganglion cells eventually put out their axons, which carry with them the specificities of the parent cells, selective and appropriate synaptic linkages are established by the action of differential affinities between matching pairs or retinal and tectal cytochemical specificities.

A main point of this hypothesis (of neuronal specificity) is that the individual specificities of the various cells, once set up, are thought to be long-lasting, or permanent. In this way the hypothesis, although aimed at the phenomena of neural development, could also account for the selective re-establishment of retinotectal connections that may occur following optic nerve lesions in lower vertebrates.

The hypothesis received its strongest support in 1963 with the main publication of the experimental results of the work of Attardi and Sperry. Previous to this, most of the experimental work on regenerating optic nerves had dealt with the re-establishment of complete maps (Gaze, 1970). And while such experiments revealed the degree of precision with which the map could be restored, they did not throw much light on the mechanisms involved. To obtain insight into how the system works it was necessary to perturb the mechanism in a controlled fashion and attempt to correlate the resulting structural abnormalities with the nature of the perturbation causing them. This is what Attardi and Sperry (1963) did.

These authors severed the optic nerve in a series of young adult goldfish and at the same time removed part of the retina on the same side. Then, after allowing some 20 days for the fibres from the remaining part of the retina to regenerate, they studied the distribution of the terminal arborizations of the optic fibres across the tectum. Following a variety of retinal lesions, they found that in all cases the fibres from the residual retina ended up in the proper, normal parts of the tectum. Particularly significant from the point of view of chemospecificity were two results: When the temporal half of the retina had been removed, the fibres from the remaining nasal half-retina, having entered the tectum rostrally in the usual way, passed right across to the caudal half of the tectum before arborizing (Fig. 2.1), and when

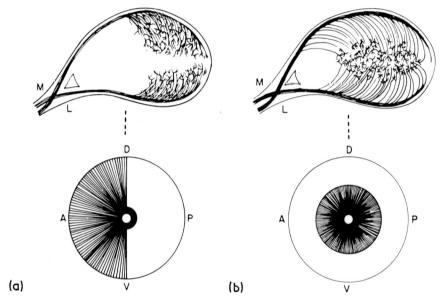

Fig. 2.1 Initial projections from retina to tectum in goldfish, after section of the optic nerve and removal of part of the retina as shown. (a) After removal of the posterior half of the retina, fibres from the residual anterior half bypass rostral tectum and arborize in caudal tectum. (b) After removal of a peripheral annulus of retina, fibres from the remaining central retina bypass peripheral tectum and arborize in central tectum. From Attardi and Sperry, 1963.

a peripheral annulus of retina had been removed, fibres from the remaining central retina coursed inwards from the two branches of the optic tract at the edges of the tectum, to arborize only when they reached central tectum.

   These last two results (together with the results of tectal graft translocation discussed later) comprise the strongest evidence yet obtained for the existence of differential affinities between pre- and postsynaptic neurons. In each case the regenerating optic nerve fibres were presented with a whole series of denervated tectal neurons with which they could have synapsed; and in each case the fibres continued past these neurons until they reached the proper part of the tectum, where they arborized. To suppose that the fibres were seeking their appropriate chemospecific partners seems to be the simplest explanation for this remarkable behaviour.

2.3 DIFFICULTIES FOR TARGET-AFFINITY MECHANISMS

In the same year that the work of Attardi and Sperry was published, there appeared also the first of a series of reports (Gaze *et al.*, 1963) that have thrown the field of specificity studies into considerable confusion. This confusion is only now beginning to be resolved. The experiments which led to these difficulties were:

(a) Studies on the connections formed by 'compound eyes' in *Xenopus*;
(b) Studies involving expansion or compression of the visual map in retinotectal mismatch situations, and
(c) Studies on the actual growth of the retinotectal connection pattern.

In 1963, Gaze *et al.* reported on the connections formed by compound eyes in *Xenopus*. Compound eyes are formed by operation on the embryo at tail-bud stage, when the polarity of the developing eye is already established. To make a double-nasal (NN) compound eye, the temporal half of one eye is removed and replaced by the nasal half of an eye from the opposite side of another embryo (Fig. 2.2).

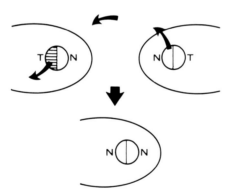

Fig. 2.2 Operation to form a double-nasal compound eye in *Xenopus*.

In a comparable fashion we can make eyes that are double temporal (TT), double ventral (VV), double dorsal (DD), as well as a variety of other combinations. Such operated eyes usually go on to develop into eyes that are apparently normal to casual inspection, except that VV eyes tend to have two 'ventral' fissures instead of one, while DD eyes may have none.

The tectal connections that are later formed by these eyes can then be studied by electrophysiological recording of the visuo-tectal map; and while a lot is now known about the various kinds of compound eye, I confine myself here to the properties of the projections from NN and TT eyes, both for reasons of space and because the principles I am considering are brought out clearly by studies of these preparations.

In a normal *Xenopus,* the greater part of the retina projects over the entire dorsal surface of the tectum; nasal retina projects caudally and temporal retina projects rostrally. In a NN or TT eye, each (similar) half of the eye projects over the entire extent of the dorsal tectal surface that would normally receive input from the whole naso-temporal extent of the eye; and the polarity of the projection from each half of the compound eye is normal for that half-eye. Thus we obtain a double projection from the compound eye with each tectal point receiving input from two positions of the retina, arranged mirror-symmetrically about the vertical meridian of the eye (Fig. 2.3).

This typical compound eye projection pattern immediately raises problems, of which the most obvious is: if the half-eyes that go to form the compound eye persist as half-eyes until recording, and if the tectum is normal, then the affinity pattern between the eye and tectum has changed. The half-eye now projects across the whole tectum. I take this point up again later; here we can note that it raises a difficulty for the original specificity hypothesis.

A variation on the retinotectal mismatch experiment of Attardi and Sperry (1963) was performed by Gaze and Sharma (1970). Instead of removing half the retina, these authors removed half the tectum in adult goldfish and found that, some months later, and irrespective of whether or not the optic nerve had been cut, electrophysio-logical mapping showed that the projection from the entire retina was now to be found compressed, in an orderly fashion, and correctly oriented, on the residual half-tectum (Fig. 2.4).

This result, confirmed many times since (Cook and Horder, 1974; Yoon, 1971; 1976; Sharma, 1972) again provides a problem for the hypothesis of neuronal specificity. If goldfish retinotectal connections are established by long-lasting retinal and tectal specificities, acting on the basis suggested by Sperry (1943), and strongly supported by Attardi and Sperry (1963), it seems impossible to account for the fact that a whole retina, which normally projects over a whole tectum, can be induced to compress its projection over only half a tectum.

This experimental result was soon augmented by others equally troublesome. Thus, Horder (1971) and Yoon (1972) showed that a surgically formed half-retina in goldfish could expand its projection to cover the whole textum. This finding (and, in principle, the map compression result of Gaze and Sharma, 1970), seemed to be in contradiction to the earlier work of Attardi and Sperry (1963). Both sets of investigations studied the results of removal of half of the retina. Attardi and Sperry (1963) showed that the projections from the residual retina went only to the appropriate half of the tectum, thus confirming the prediction of the hypothesis of neuronal specificity, whereas Horder (1971) and Yoon (1972) showed that the projection of the residual retina was spread over the whole tectum. These

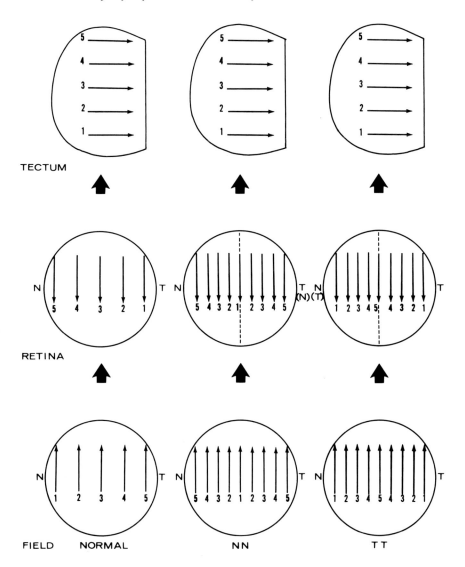

TECTUM

RETINA

FIELD   NORMAL   NN   TT

Fig. 2.3 Retinotectal projections from double-nasal and double-temporal compound eyes in *Xenopus*. The numbered arrows on the tectal diagrams represent rows of electrode positions. Corresponding rows of positions on the retinae and on the visual field charts are indicated below.

contradictory results led to various suggestions about possible differences in the mode of surgical operation; but the matter remained unresolved until it was suggested (Sharma, 1972; Gaze, 1974) that the difference between the results was

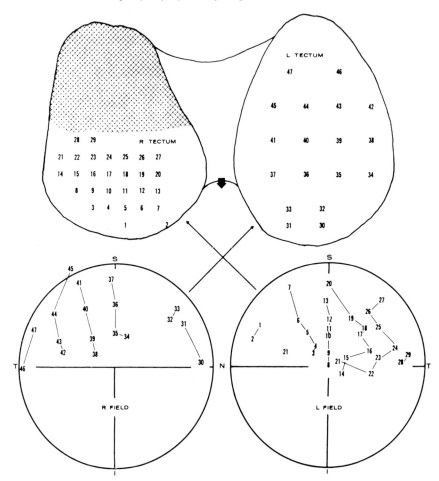

Fig. 2.4 Compression of the entire upper field of an eye onto a residual rostral half-tectum in goldfish. Numbered positions on the tectal diagram represent electrode positions. Corresponding numbers on the field charts indicate optimal stimulus positions. The projections from the right field to the left tectum is normal; that from the left field to the right tectum is compressed. From Gaze and Sharma, 1970.

a question of timing. Attardi and Sperry (1963) had studied the residual projections about 20 days after operation and found a truncated, but otherwise strictly normal retinotopic projection; Gaze and Sharma (1970) had mapped their animals about three months after operation and found field compression; and Yoon (1976) showed that 40 days or more were required, after partial tectal ablation, before a compressed map could be found. It seemed, therefore, that, had Attardi and Sperry waited longer, they would have observed changes in the projection: the residual half-retina would have expanded its projection across the entire tectum. This has since been shown to occur (Cook and Horder, 1974; Schmidt *et al.*, 1977).

A further result has been reported which adds yet more difficulty for the neuronal specificity hypothesis as originally proposed. Horder (1971) and Yoon (1972) have shown that a half-retina in goldfish can project, in properly ordered and oriented fashion, over the *inappropriate* half-tectum (Fig. 2.5).

Up until very recently all studies on the formation of retinotectal connections had been performed on adult animals or juveniles; and thus on regenerating, rather than developing systems. However, the hypothesis of neuronal specificity was intended to account for neural development primarily, and regeneration only secondarily. It therefore seemed a good idea to study retinotectal projections *as they developed* rather than to concentrate solely on regeneration and then infer backwards to the events of neurogenesis. *Xenopus* is an ideal animal for this sort of approach, because of its very protracted periods of retinal and tectal development; and electrophysiological observations of the developing retinotectal projection have been described by Gaze *et al.*, (1974).

The retina in *Xenopus* grows from the optic nerve head outwards, by the serial addition of cells (in all retinal layers) at the ciliary margin and this cellular addition continues until after metamorphosis (Straznicky and Gaze, 1971). The tectum grows from about stage 40/45 until metamorphosis (or even later) but in this case cell addition occurs from rostro-lateral to dorso-medial (Straznicky and Gaze, 1972). The earliest visual responses have been recorded in the tectum at stage 43 and synaptic transmission has been demonstrated from stage 45 (Chung *et al.*, 1974); from shortly thereafter an ordered retinotopic map is to be found and this moves caudally across the tectum as both tectum and retina grow (Gaze *et al.*, 1974). The continual shift of the field map across the developing tectum, together with the continual arrival at the rostral pole of the tectum (the first part to develop) of new retinal fibres from the growing temporal margin of the retina, led to the suggestion (Gaze *et al.*, 1974) that, during the establishment of the map, retinotectal synaptic connections were undergoing progressive shifts caudalwards. A comparable suggestion has been made in relation to the visual system of the goldfish, where retina and tectum appear to continue growth throughout life (Johns, 1976).

If these suggestions hold up, and this seems likely (Scott and Lázár, 1976; Gaze, Keating and Chung, in preparation), despite recent comments to the contrary (Jacobson, 1976) they will require modification of the neuronal specificity hypothesis, since the tectal connections made by a particular retinal cell early in development need not be the same as those made by the same retinal cell later in life (Fig. 2.6).

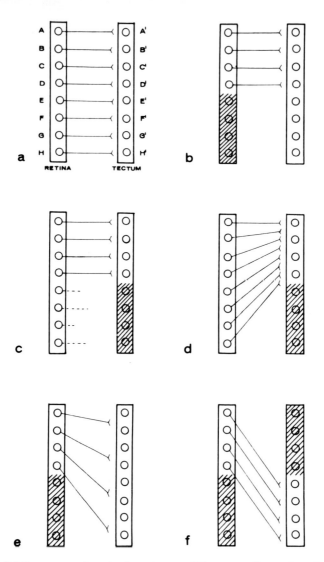

Fig. 2.5 Summary of projection abnormalities seen after various forms of surgical interference with the goldfish visual system. In each diagram the retina and tectum are shown as schematic arrays of neurons, carrying the place-related specificicity labels indicated by the capital letters. (a) Normal retino-tectal projection; (b) Initial projection of half-retina, as found by Attardi and Sperry, 1963; (c) Initial projection of retina to half-tectum (Gaze and Sharma, 1970; Cook and Horder, 1974); (d) Later projection of retina to half-tectum (Gaze and Sharma, 1970; Yoon, 1971); (e) Expanded projection from a half-retina (Horder, 1971; Yoon, 1972; Schmidt *et al.,* 1977); (f) Translocation of half-retinal projection to inappropriate half-tectum (Horder, 1971; Yoon, 1972). From Gaze, 1974. Reproduced by permission of the Medical Department, The British Council.

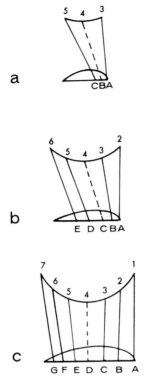

Fig. 2.6 Retinotectal relationships along the naso-temporal axis of the retina and the rostrocaudal axis of the tectum, at various stages (a) early tectal innervation; (b) mid-to-late larval life; (c) post-metamorphic) during development in *Xenopus.* The letters (tectal) and numbers (retinal) represent actual cellular elements and not specificities. Because of the differing modes of growth of retina and tectum, the retinotectal projection appears to move caudally across the tectum during development. In each diagram, most-nasal retina is to the left and most-rostral tectum is to the right; the dashed line represents the projection from central retina. From Gaze, 1974. Reproduced by permission of the Medical Department, The British Council.

2.4 RIGIDITY VERSUS PLASTICITY OF CONNECTIONS

With the publication of these various results there has grown up a protracted controversy as to the nature of the connectivity changes observed, and the nature of the causal mechanisms. The experiments mentioned in the previous section (compound eye studies; contraction/expansion phenomena; the shift of the visuo-tectal map during development) indicate that retinotectal connections are not unmodifiable. They can alter; but they do so in an ordered and properly polarized fashion.

In the face of this evidence concerning the mobility of certain neuronal connections it seemed that, to encompass for instance the compression of a retinal projection onto a half-tectum, there were two possibilities within the general framework of ideas on neuronal specificity (Fig. 2.7):

(a) If the mapping function which represents the affinity of particular retinal cell labels for particular tectal cell labels remains constant, then the cells of a residual half-tectum must change their labels, to ensure that there is now a complete specificity-structure over a diminished array of cells. Or,

(b) If the cells of a residual half-tectum retain their cytochemical specificities, then the mapping function must be different. It cannot simply be the one-to-one matching of retinal and tectal labels that was originally envisaged, but must in some sense be contextual.

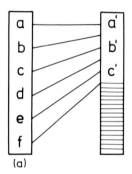 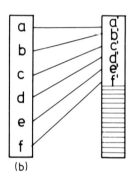

Fig. 2.7 Two interpretations of retinotectal compression. In each diagram retina is on the left and tectum on the right. Cellular specificities are indicated by letters. (a) If the tectal specificity labels remain unchanged after tectal surgery, the mapping function relating retina to tectum can no longer involve the linkage of matching labels. (b) Preservation of a mapping function involving linkage of matching labels requires that the labels should change.

We have argued previously (Straznicky *et al.*, 1971) that, in any such mapping between two neuronal arrays, there are three variables to be considered: The array of retinal cell specificities; the array of tectal cell specificities and the rule, or mapping function relating the one to the other. It is conceivable that the retinal specificities, or tectal specificities, or both, might change following surgical intervention. Such changes would then be akin to the phenomenon of embryonic 'regulation' following injury; but in the present situation occurring in an adult animal. Furthermore, there seems to be no reason why the mapping function itself might not give different results in different circumstances; that is, behave in a contextual fashion. It can be seen that there are great difficulties in making firm statements about any of the three parts of the systems, since assessment of one part depends upon the assumed stability of the other two. These matters are considered in some detail elsewhere (Gaze and Hope, 1976); and some of the evidence relating to regulation is discussed in a later section.

2.5 REASONS FOR THE PRESENT CONFUSION OVER SPECIFICITY

There seem to be two reasons for the present confused state of the specificity field: Firstly, the situation being dealt with, the formation of nerve connections, is more complicated than originally seemed to be the case; and secondly, the hypothesis of neuronal specificity has never been articulated in detail. This lack of detailed articulation has permitted different investigators to maintain conflicting views as to which experimental results were, or were not, compatible with the hypothesis. The level of confusion so engendered is well illustrated by two recent papers. Thus we read, in relation to retinotectal mismatch experiments,

". . . surgically created contexts never faced by normal retinal ganglion cells and tectal cells may have only limited relevance for the theory . . . these points are important, surely; but the theory simply does not address them" (Hunt and Jacobson, 1974). Conversely, in a paper published by Meyer and Sperry in the previous year, we find,

". . . Given stabilized tectal specificities, a lasting behavioural hemianopia would be predicted to follow hemitectal ablations according to the original explanation with no compression of the visual field measured electrophysiologically" (Meyer and Sperry, 1973).

These fundamental disagreements over what is meant by the hypothesis of neuronal specificity follow directly from the rather general terms in which the hypothesis was couched. Thus it was proposed (Sperry, 1943; 1944; 1945; 1951; 1963; 1965) that retinotectal connections form under the influence of differential affinities between retinal axons and tectal cells. Although such a formulation is satisfactory for outlining the general class of events that is envisaged, in this form it is so general as to be virtually untestable. Because such a general statement says nothing about how the affinities may work — whether they are to be thought of as mosaic, non-interactive affinities, where the behaviour of any two affine units is uninfluenced by anything going on elsewhere in the whole system, or whether the affinities are in some sense contextual and interactive — the general statement makes no predictions about the results that might be obtained in situations such as the retinotectal mismatch experiments.

Another major cause of difficulty is the ease with which it is possible to confuse specificities with affinities. The hypothesis proposes that cytochemical specificities come to distinguish each cell in each array. The cells are then linked by the action of selective affinities. That is, the affinities *are* the mapping rule or function. According to how one defines the affinities, or mapping function, one can obtain different patterns of connection in different circumstances; but the specificities, i.e. the individual cytochemical labels on each cell, need not have changed. This requirement for the investigator to take into account both sets of specificities and the mapping function, or affinities, is very important as I have argued earlier.

It is now fourteen years since the first reports of reduplicated maps from compound eyes in *Xenopus* (Gaze *et al.,* 1963) and seven years since the first report of field compression onto a half-tectum (Gaze and Sharma, 1970). It is perhaps surprising that, over all this time, there has been little attempt to refine and extend the useful ideas of neuronal specificity. The result has been that, particularly in the last three years or so, experiments of the mismatch type have reached levels of complexity which make them difficult to assess in terms of the original hypothesis of neuronal specificity. A new approach is obviously needed; and this new approach is not necessarily a change in the type of experiments that are being performed, for these continue to produce immensely valuable results, but rather more and better theoretical work. In this way we may hope both to extend the usefulness of Sperry's concepts and to help in the design of more searching experiments.

A start has been made in this direction. In the last two years three papers have appeared dealing with theoretical aspects of retinotectal connectivity. These are by Prestige and Willshaw (1975); Hope, Hammond and Gaze (1976) and Willshaw and von der Malsberg (1976). The first of these I consider here, and the latter two are discussed in a later section.

The work of Prestige and Willshaw (1975) is one of the most important papers to appear in the field of specificity studies, as well as the first to tackle seriously the theoretical background of affinity mechanisms. These authors investigated, by computer simulation, the working of various models based on affinities; and what makes their work particularly interesting is that they considered affinity from the points of view of the pre-synaptic fibre and the postsynaptic cell, separately.

Perhaps the most widely held view of affinity (in the nervous system) has been that it is an interactive, or two-way phenomenon. This is clearly the view implied by Sperry's usage of the term. An alternative way of looking at affinity may be illustrated by way of an analogy:

Consider two sets of people who are going to form linkages. The affinities which determine the linkages formed may be of two types. In the one case we may represent the affinities by use of complimentary markers in the two populations. For instance, colours of clothing may vary among the first group of people, with matching colours among the second group. Then blue would link with blue, green with green and so forth. Such an affinity mechanism (which would be comparable to a lock and key molecular mechanism) is interactive in the sense that two persons, one from each group, have to get together and agree before linkage can occur.

Consider another case, where the people in each group have names rather than colours. We can call the first group A, B, C etc and the second group 1, 2, 3 and so on. Person A has to choose one of the opposite set; and he may indicate relative preferences. Suppose A chooses in the order of preference 1, 2, 3. Then B and C have to choose. They can choose in the same order as A, but they need not. B could choose 2, 3, 1: and C could choose 3, 1, 2. If B and C choose in the same order as A, then person 1 becomes the most sought after member of the opposite set, by common consent. But now the persons in the second set are allowed to make their choices from among the members of the first set. Again, each member of the second set can choose the members of the first set in any order of preference. Obviously, the number of possibilities increases with the number of people in each set; and so does the liklihood of conflict of interests between the two sets.

This type of affinity system includes the Sperry type as a special case. If A chooses $1_{2,3}$, B chooses $2_{1,3}$ and C chooses $3_{1,2}$, while 1 chooses $A_{B,C}$, 2 chooses $B_{A,C}$ and 3 chooses $C_{A,B}$, then we have what Prestige and Willshaw (1975) call 'direct' or 'rigid' matching, which seems to represent the situation in Sperry's original hypothesis. As Prestige and Willshaw point out, we must assume that, in a continuous mapping of one set onto another, the labels indicating position in the sets are graded in quality, with a single gradient of labels representing a single axis of space. The alternative would involve the existence of a 'supermap' to interpret discontinuous labels in terms of graded positional information; and this seems both inelegant and unnecessary. The assumption that postional labels are graded would mean (as Sperry (1943) suggested) that labels on adjacent cells would differ from each other less than labels on cells that are further apart. This allows us to introduce any desirable amount of 'slop' into the matching mechanism, by adjusting the sharpness of the affinity function. By making the affinity function very narrow we can ensure that the fibre has maximal affinity for one particular cell and very little for any other; or, if we make the function curve wider, we can ensure that the maximal affinity is merely a slight preference for one cell, with affinity for near neighbours not much less (Fig. 2.8).

Prestige and Willshaw (1975) give the name Group 1 Mechanism to cases of direct or rigid (Sperry type) matching. Here, an axon j has a maximum or near maximum affinity for member j' of the postsynaptic set, and less for all other members. And conversely, cell j' has maximum affinity for axon j of the pre-synaptic set. In such a mechanism, apart from the possibility of regulation, cells make contact independently of their fellows and are not much affected by removal of part of the set. Such a mechanism, thus, will account very well for the results of Attardi and Sperry (1963) but will require regulation of cell labels to permit compression or expansion.

The interesting new variety of affinity mechanism introduced by Prestige and Willshaw is what they call the Group 2 Mechanism. Here, all axons have maximum affinity for making and retaining contacts at one end of the postsynaptic set of cells, and progressively less for cells at greater distances from that end (Fig. 2.8). Similarly, all postsynaptic cells have maximum affinity for axons from one end of the pre-synaptic set, while axons remote from this end have correspondingly less likelihood of retaining any contacts. There is thus a graded affinity within both pre- and post-synaptic sets and the contact behaviour of individual cells and axons will now be highly dependent on the presence or absence of their neighbours.

In Group 2 Mechanisms, simple affinity gradients which in each set decrease monotonically from one end of the set, do not produce a map. The easiest way to introduce mapping into the system is by a form of competition. If the number of branches per axon and the sites per postsynaptic cell are limited, then high affinity

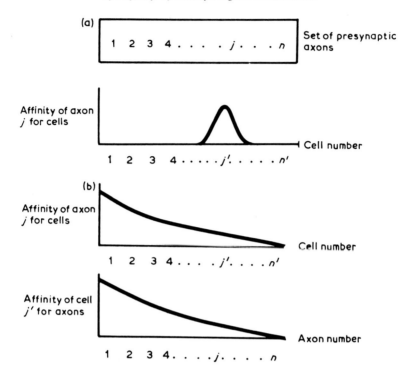

Fig. 2.8 Affinity distributions in Group 1 and Group 2 mechanisms. (a) Example of the affinity preferences of an axon j working according to a Group 1 mechanism. Axon j has peak affinity for postsynaptic cell j'. A characteristic affinity curve of a similar type if required for each axon and postsynaptic cell. (b) Examples of the affinity preferences of an axon j and a postsynaptic cell j' working according to a Group 2 mechanism. Each has peak affinity, not for its opposite number, but for the cell or axon at the top end of the gradient. From Prestige and Willshaw, 1975.

elements rapidly become saturated, enabling lower affinity elements to establish contact because they have nowhere else to go. Prestige and Willshaw show that such a Group 2 Mechanism will produce a proper map, will account for expansion of each half of a compound retinal projection (without alteration of retinal labels) and also for the progressive shift of contacts that seems to occur during tectal development. However, a Group 2 Mechanism, by itself, will not produce expansion or contraction in a mismatch experiment. To obtain this type of result, one has to introduce a further constraint such as contact equalization — allowing the total number of contacts formed to return to the original number — and this could be described as the form of regulation, of synapse number rather than cell labels.

A great virtue of Prestige and Willshaw's approach is that, for the first time, some of the consequence of various affinity assumptions are investigated. It is possible, however, to devise mapping mechanisms which do not involve any retinotectal affinities whatsoever. Some of these are discussed in the following section.

2.6 TIME-POSITION MECHANISMS

Here we envisage that the formation of neuronal connections is a direct result of the precise genetic control of the time of differentiation of the nerve cell from its precursor. This, together with the position and orientation of the cell in the organism, will automatically mean that each neuron develops within a particular and unique micro-environment. This local environment would then constrain the outward growth of the axon by mechanisms which are well recognized, although not properly understood, such as contact-guidance. In a pure time-position mechanism, no form of neuronal chemospecificity is required.

It may be seen that time-position mechanisms place a great weight of responsibility on genetically programmed simultaneity of events. Furthermore, various ancillary factors would have to be assumed, such as that a target neuron would be prepared to receive an axon contact only if the target was in a suitable state of differentiation – another requirement for timing. And we could add dendritogenic stimulation contingent on the arrival of an axon.

In a simple system, involving few neurons, we can imagine that such a mechanism could work. In a large, many-celled and complex system we may well have doubts on the adequacy of the mechanism. This is partly because, in a small system with few elements, it is easy to see how the various connection-events could be adequately spaced out along the time scale, while still maintaining the overall period allocated to neural development within reasonable bounds. With a large, many-celled system, the various connection events could not be spaced out in a single distribution along the time axis since the interval between the events would rapidly tend to zero. Even if the system is thought of as being sub-divided into various domains separated in space, so that the timing of the events in one domain need not clash with the timing of comparable events in another domain, the mechanism still seems, with increase in cell number, to become rapidly too unwieldy to work. Furthermore, since time is a single dimension, differential timing of axon outgrowth could not, by itself, account for the generation of maps ordered in three dimensions.

It would, however, be pointless for us to abandon time-position mechanisms entirely, just because we cannot see how they could account for the whole development of connectivity in a large system. We know that different neurons in the CNS do differentiate at different times. There is a rostro-caudal time gradient of differentiation of neurons throughout the neural axis; and this is particularly well shown in the development of the optic tectum (Straznicky and Gaze, 1972; Lázár, 1973). Furthermore, even in complex systems it seems always to be the case that fibre outgrowth itself follows a spatially related time-course. To an extent this would follow from the known differential timing of neuron maturation; and the timing differences of fibre outgrowth are particularly well seen in the amphibian retina where, since retinal development is extended over a period of months, we are able to observe events as if they were slowed down in a cine film.

The strongest evidence in favour of time-position mechanisms up to the present time comes from the study of certain invertebrate nervous systems, in particular the visual system of *Daphnia magna* (Lopresti *et al.*, 1973). However, as has been argued before (Gaze and Hope, 1976), all such evidence to date is permissive rather than compelling. The observations on these invertebrate systems could perhaps be accounted for in this way; but this is not strong evidence since no analytical experiments, such as could exclude time-position mechanisms, have yet been performed.

As I shall show later, my present view is that a form of time-position mechanism could indeed have relevance to the initial development of the amphibian visual system; but it cannot, by itself, account for all the phenomena of visual development, nor for the phenomena of optic nerve regeneration.

2.7 FIBRE-SORTING MECHANISMS

Other types of mechanisms may exist which could produce ordered maps and which would not require target-affinity or rely upon time-position. One class of these we call fibre-sorting mechanisms (Gaze and Hope, 1976). We have previously described such a model in relation to the formation of retinotectal connections in the goldfish (Hope, Hammond and Gaze, 1976), and have shown, by computer simulation, that, provided that the fibres can sort themselves in relation to their retinal provenance and to tectal polarity, a normal retinotectal map can be established in the absence of any form of target-affinity (Fig. 2.9). This mechanism, which we call the 'arrow model', since the only help offered by the tectum to the incoming fibres is to provide local polarity information which can be thought of as arrows on the surface of the tectum, can also account for various other manifestations of optic nerve regeneration in goldfish. Thus the model will give rotation of the part of the retinotectal map covering a rotated tectal graft (Fig. 2.10); it will permit compression of a retinotectal map over a half-tectum; and it will give expansion of a retinal map over an enlarged tectum.

Since all these phenomena, previously taken as manifestations of differential affinity, or alterations of affinity, can be produced by computer simulation based on the rules of the arrow model, which does not use affinity at all, we can ask how good the evidence for differential affinity actually is; and the answer must be: still very good. The critical result for the demonstration of the existence of affinities is *not* the formation of a normal map in regeneration, the formation of a rotated map over a rotated tectal graft, or the production of map compression/expansion phenomena. The result that is most useful is that obtained in the retinotectal mismatch experiment by Attardi and Sperry (1963), or the formation of a map over a tectum bearing a

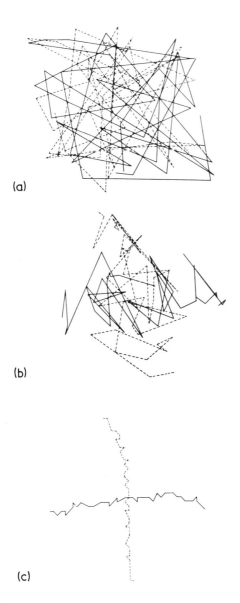

(a)

(b)

(c)

Fig. 2.9 The formation of a retinotectal map on the basis of the arrow model. Computer simulation of retinotectal mappings according to the simple-arrow model. (a) Tectal mapping of a horizontal (full line) and a vertical (dotted line) row of retinal positions, each passing through the centre of the retina. The diagram shows the initial configuration, which is a pseudo-random mapping. (b) As in (a) but after 40 iterations of the interchange programme. Sorting of the positions is already evident. (c) As in (a) but after 60 iterations of the interchange programme. The map is sorted. From Hope *et al.*, 1976.

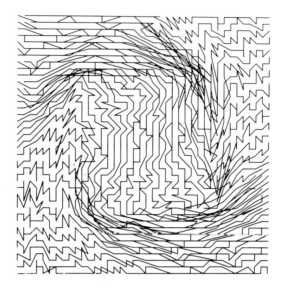

Fig. 2.10 Retinotectal map according to the arrow model in the case of a rotated tectal graft. The diagram shows a computer-simulated tectal mapping of the complete array of 40 rows of retinal positions parallel to one axis. In this case the central part (20 x 20) of the 40 x 40 tectal array was rotated by 90° and the result was obtained after 60 iterations of interchange programme. The map is rotated within the area of the graft and those positions that are appropriate to the graft have ended up within it. From Hope *et al.*, 1976.

translocated graft. In the latter case, since in the arrow model the only information on the tectum is local polarity, if we translocate grafts rostro-caudally, without graft rotation (Fig. 2.11), the incoming fibres should not be able to tell the difference and a normal map should result. In fact, translocated tectal grafts give correspondingly translocated parts of the retinotectal map (Hope, Hammond and Gaze, 1976) and this is one of the strongest arguments in favour of an affinity mechanism.

Thus the arrow mechanism can be ruled out as being inadequate to account for these regeneration phenomena. As I shall show later, however, there is still a good chance that some such fibre-sorting type of mechanism may be involved in neurogenesis during early life.

2.8 OTHER MECHANISMS

Two models have recently appeared which require mention, on account of both their elegance and their biological likeness. Both models have been proposed by Willshaw and von der Malsberg (1976; and von der Malsberg and Willshaw, in preparation).

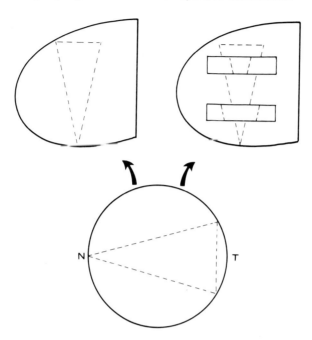

Fig. 2.11 Translocation of tectal grafts. The visual field chart (lower diagram) has a triangle superimposed on it to enable various parts of the field to be distinguished. N, nasal; T, temporal. The tectal diagrams (above) show the projection of the visual field triangle on a normal tectum (left) and on a tectum bearing a rostro-caudally translocated graft. See Hope *et al.,*1976.

In their first model Willshaw and von der Malsberg (1976) make use of the fact that, in continuously ordered mappings, neighbourhood relationships are preserved: The prospective targets of neighbouring pre-synaptic cells are themselves neighbours in the postsynaptic array. On this basis the authors propose a model for the form-ation of continuously ordered maps which works by self-organization, using some of the electrical properties of neurons.

The model is made of two independent parts. One part ensures that, as a result of an optimizing process based on near-neighbour electrical interactions and modifiable synapses, neighbouring pre-synaptic cells come to connect with neighbouring post-synaptic cells. As the authors point out, if this condition is met by all pairs of neighbouring cells, the resulting mapping will be a topographical one. The other part of the model sets boundary conditions which determine the size and orientation of the map. A map based on neighbourhood-preserving mechanisms only, without boundary conditions, would not be very useful since, although the map would be correct in its internal order, it need have no particular orientation or size. The authors provide orientation information by postulating the existence of polarity

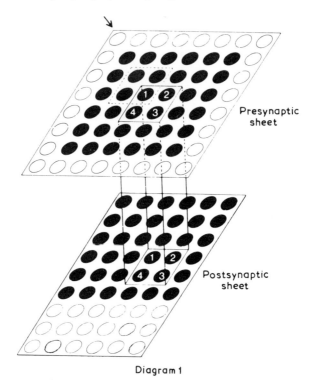

Diagram 1

Fig. 2.12 The formation of a continuously ordered neural map according to an electrical self-organizing mechanism.

Diagram 1. The two sheets which are to inter-connect. Filled circles represent the members of the two 6 x 6 sheets involved in the mapping in diagrams 2a and 3. The pairs of polarity marker used for diagram 2a are labelled 1, 2, 3, 4; for diagram 3 the same postsynaptic polarity markers were chosen, but this time their partners are the four cells enclosed by the dotted line drawn on the pre-synaptic sheet. The cells added in the calculations for diagrams 2b, c and d are denoted by unfilled circles. The arrow indicates the corner of the pre-synaptic sheet to be placed top left in the maps of diagrams 2 and 3.

markers, cells of a particular small pre-synaptic region which initially make contacts, in the required orientation, with a small postsynaptic region (Fig. 2.12).

Willshaw and von der Malsberg (1976) point out that their polarity markers do not introduce any precise, pre-programmed cellular specificity into their model. They argue that the weak specificity required for the polarity markers contains only enough information to specify orientation; and furthermore, the chosen pre-synaptic region can make initial connections with any small postsynaptic region, not just that region to be connected with it in the final mapping.

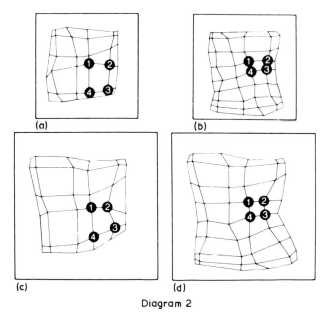

Diagram 2

Fig. 2.12 The formation of a continuously ordered neural map according to an electrical self-organizing mechanism.

Diagram 2. Maps of post-synaptic cells drawn on the pre-synaptic sheet for various model systems. In each diagram the thick black line tracing out a square denotes the boundary of the pre-synaptic sheet, oriented with the corner marked by the arrow in diagram 1 placed top left. For each post-synaptic cell the centre of the cluster of pre-synaptic cells connected to it was plotted out. Points associated with neighbouring postsynaptic cells were then joined by a straight line. Since only cluster centres were plotted, the post-synaptic maps so constructed are necessarily smaller than the appropriate pre-synaptic sheets; the half-width of the distribution of pre-synaptic cells connected to a given postsynaptic cell never exceeded $\frac{9}{10}$ of the pre-synaptic cell inter-spacing. Postsynaptic polarity marker positions are indicated as in diagram 1.

(a) The mapping between the two 6 x 6 sheets which has developed after 15 000 trials. Diagrams b, c and d show the new mappings obtained after performing various manipulations on the system in the state shown in a.

(b) The mapping obtained, after 9000 trials, when the number of post-synaptic cells have been increased from 36 to 54 by adding three rows of 6 cells, to make a 6 x 9 sheet, and the pre-synaptic sheet was unchanged.

(c) The mapping obtained, after 10 000 trials, when the 6 x 6 pre-synaptic sheet had been enlarged by adding a band of cells round the edge to make an 8 x 8 sheet, and the postsynaptic sheet was unchanged.

(d) The mapping obtained, after 15 000 trials, when both sheets were enlarged, to 64 and 54 cells respectively.

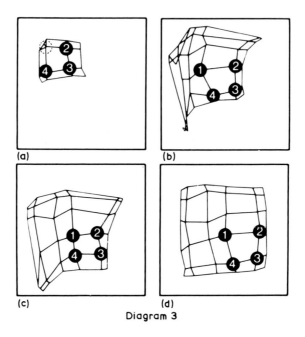

(a)   (b)

(c)   (d)

Diagram 3

Fig. 2.12 The formation of a continuously ordered neural map according to an electrical self-organizing mechanism.

Diagram 3. The development of a mapping between two 6 x 6 sheets in the case when the two sets of polarity markers occupied non-corresponding regions on the two surfaces. Conventions are as in diagram 2. Only the positions of those postsynaptic cells which had established definite contacts were plotted.

(a) By 5000 trials only the polarity markers and their neighbours had established definite contacts. The mapping set up by this stage is between cells in the top left-hand quarter of the pre-synaptic sheet and those in the bottom right-hand quarter of the postsynaptic sheet.

(b) After 20 000 trials. To accomodate the new cells which established contact, albeit in a disorganized manner, the map of the postsynaptic sheet has shifted down and to the right.

(c) By 30 000 trials postitions of all postsynaptic cells could be plotted out but the map is distorted.

(d) After 50 000 trials. A stable and ordered map is produced.
From Willshaw and von der Malsberg, 1976.

We can perhaps strengthen these statements of Willshaw and von der Malsberg. The term specificity as used in the previous paragraph is confusing because it is used in two different senses. In the first place, the authors talk of precise pre-programmed cellular specificity; and this is the sense in which the word is used by Sperry in discussion of the hypothesis of neuronal specificity (1943, 1951). In the second place, Willshaw and von der Malsberg talk of the weak specificity required to specify orientation; and this, being merely a means towards ensuring congruence of polarity between a group of pre-synaptic and a group of postsynaptic cells, need not involve cellular specificity of any sort. Cellular specificity is one way of achieving this end but there are others (Hope, Hammond and Gaze, 1976).

This 'electrical', self-organizational model of Willshaw and von der Malsberg has been shown, by computer simulation, to establish normal maps; and it also shows 'systems-matching' behaviour (Straznicky *et al.,* 1971; Gaze *et al.,* 1972; Gaze and Keating, 1972) in that pre- and postsynaptic arrays will match up irrespective of whether one is smaller or larger than the other. What the electrical model will not do is account for the results of graft translocation; or the early results of a half-retinal projection to a whole tectum, where initially a half-map is formed over the appropriate half of the tectum, and only later does the half-retinal projection spread over the whole tectum (Schmidt *et al.,* 1977). To account for these phenomena, it seems, we must invoke affinity (specificity).

This requirement is met by the second model of these same authors (von der Malsberg and Willshaw, in preparation). This model, which we may perhaps call the 'Tea Trade model', since this is the analogy the authors use to explain how the system works, is a molecular diffusion model and thus could be applicable to developmental situations where electrical spike activity may not yet be possible; and moreover the model has a memory, so that in some regeneration situations (e.g. half-retina) the axons first seek out their original postsynaptic partners and only later co-operate with one another to produce an expanded or compressed map. The essential idea of the model is that a fixed set of chemical markers labelling the pre-synaptic cells is induced, through modifiable synapses, onto the postsynaptic sheet.

The Tea Trade model has also been shown, by computer simulation, to establish normal maps. It will show systems-matching behaviour in mismatch situations, although this state is not reached immediately; initially, fibres regenerate to their original sites and later move, in an ordered fashion, to produce a systems-match. And furthermore, the model will give spreading of the connections from each half of a compound eye (Gaze *et al.,* 1963) across the entire tectal receiving surface, producing a mirror-image map; and this can occur without altering the set of pre-synaptic markers. The Tea Trade model will also account for the occurrence (at least initially) of translocated maps across translocated regions of tectum.

2.9 RECENT DEVELOPMENTS IN SPECIFICITY STUDIES

We come finally to the point of all the previous discussion. Although various attempts are now being made to account for the formation of ordered maps without invoking target-affinity, as described in the previous two sections, the most widely accepted hypothesis is still an affinity mechanism; and the recent experiments which have led to our present ideas on the development of connections have mostly been variations on the retinotectal mismatch theme. We may state at the outset that all is not now clear. Indeed, the experiments fall into two main categories: those that appear to fit a pattern and those that do not. We take the seemingly inconsistent experiments first.

Sharma (1977) reports experiments on goldfish where the temporal half of each retina was removed and each residual nasal retina was allowed to expand over its contralateral tectum. Five months after removal of the half-eye one tectum was ablated and the projection to the remaining tectum were mapped 278 days later. The contralateral half-eye still projected in expanded fashion over the whole tectum; the ipsilateral half-eye, however, (induced by removal of its own tectum to innervate the other), only covered, in properly polarized order, the rostral, incorrect, half of the tectum. By 40 days later, the ipsilateral half-eye had also expanded its projection across the whole tectum.

Sharma (1975) has described the retinal projection to surgically-created 'compound' tecta in adult goldfish. Both double-rostral and double-caudal tecta were investigated. Initially, one month after operation, the compound tectum showed a reduplicated projection coming only from the appropriate half retina. By three months after surgery, other animals showed a restored normal field map over the whole (normal half plus transplanted half with reversed rostro-caudal polarity) compound tectum. In these experiments, therefore, the transplanted half-tectum appears to have become re-polarized; either by the residual intact half of the tectum or by the incoming optic nerve fibres. As will be seen later, this particular result could perhaps be classified with those of the second group.

Cronly-Dillon and Glaizner (1974) report experiments where a normal goldfish eye was induced to innervate the ipsilateral tectum, the eye contralateral to which had had its temporal half-retina removed. A year later the visual map on the innervated tectum was found to come from the temporal half-field (nasal retina) of each eye; and instead of both expanding over the whole tectum they sub-divided the tectum such that the ipsilateral half-projection was to the caudal tectum while the contra-lateral half-projection was to the rostral, inappropriate, tectum.

These peculiar experimental results cannot, at present, be explained. One question they raise is under what circumstances the pre-synaptic fibres may be considered to form a single population. The relevance to map formation of defining cell and fibre populations has been discussed in a previous paper (Gaze and Hope, 1976).

The next group of experiments to be considered form a more coherent story and deal particularly with the question of whether or not a regulation of retinal or tectal specificities occurs in the case of a mismatch situation giving expansion or compression of the map.

It has been argued (Meyer and Sperry, 1974) that, in the compound eye experiments on *Xenopus,* each half of a compound retina may regulate its specificities to form a complete set. However, while repolarization of one eye fragment by the other is known to occur in some circumstances (Hunt and Frank, 1975), regulation of cell labels in compound eyes probably does not. In *Xenopus,* ventral retinal fibres destined for dorsal tectum enter via the medial branch of the optic tract, while dorsal retinal fibres destined for lateral tectum enter via the lateral branch of the tract. Animals with a double-ventral compound eye are found to have no lateral tract (light microscopy, silver staining) and an enlarged medial tract (Fig. 2.13; Horder, Straznicky and Gaze, unpublished). Thus fibres from each (ventral) half of the retina took the normal path for ventral retinal fibres. If, following operation, both halves of the retina had regulated to provide each with a full dorsoventral specificity structure, we would expect both tracts to be normal.

The compression of the projection from a normal retina onto a half-tectum, and the expansion of the projection from a half-retina over a whole tectum, have been attributed to regulation of specificities in half-tectum and half-retina respectively (Meyer and Sperry, 1974). There is now a considerable amount of evidence, however, which indicates that compression and expansion phenomena are not the result of regulation.

Any regulation phenomenon requires that the tissue undergoing regulation should have new boundaries established for it; and the regulation then occurs within the limits of the new boundaries, to re-establish to totality of the pattern. One cannot merely draw an imaginary line across the tissue in a particular place and announce that this is to be the new boundary. The boundary has to have physical reality; and indeed, it is usually established by surgical truncation of the tissue involved. While the nature of the mechanisms underlying regulation is not known, the phenomenon has been extensively discussed in the context of ideas about positional information (Wolpert, 1971).

Several recent experiments indicate that one can obtain field compression on the tectum in the absence of a complete new boundary system. For instance, when two grafts are translocated rostro-caudally on the goldfish tectum, apart from the corresponding map translocation that occurs (Hope, Hammond and Gaze, 1976), there is often to be seen a field compression over the 'normal' tectal tissue between the grafts (Gaze and Hope, in preparation). The compression here is only found on that region of tectum bounded by the caudal edge of the rostral graft and the rostral edge of the caudal graft. The compressed region is in direct continuity with the normal regions of tectum both laterally and medially.

Compression in the anuran tectum has also been demonstrated in the absence of a complete new boundary system. Freeman has shown (1977) that the application of α-bungarotoxin to a small square region of the tectum causes the map to disappear from this region; and the displaced map positions are relocated, in compressed but proper order, just outside the border of the treated region of tectum (Fig. 2.14). The displaced positions give both pre- and postsynaptic activity. This experiment thus shows the establishment of localized compression in a situation where regulation cannot be invoked.

The development and mechanism of the expansion of a half-retinal projection in

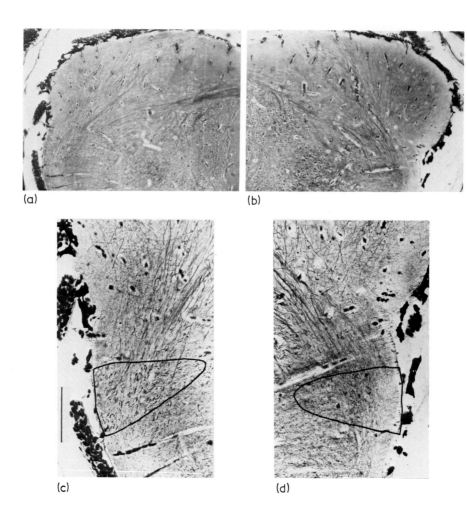

Fig. 2.13 The optic tracts in *Xenopus* with one double-ventral eye.

(a and b) Coronal sections at the level of the rostral tectal margins, to show the positions of the lateral and medial optic tracts on each side.

(c) Lateral tract opposite the normal eye.

(d) Position of lateral tract opposite the compound eye; no tract fibres seen. 15 μm sections, Holmes' silver stain. Bar = 100 μm (lower photographs). From Gaze, Horder and Straznicky, unpublished.

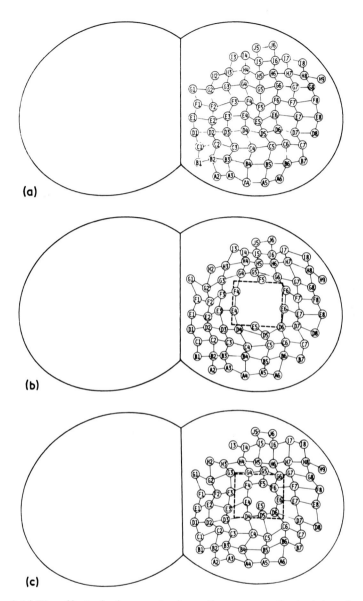

(a)

(b)

(c)

Fig. 2.14 The effect of α-bungarotoxin on the anuran retinotectal projection. The figure shows plastic changes in position of class 4 terminals following localized application of α-bungarotoxin to normally innervated tectum in *Bufo marinus*. Control map shown in a. In b, obtained one week after applying a square of millipore filter soaked in α-bungarotoxin to the region indicated by the dashed lines, the terminals previously located there (e.g. E4 to E6, F4 to F6, G5) have shifted position and compressed into regions near the borders, displacing terminals previously located there. In c, obtained two weeks later, terminals have essentially returned to their original position. From Freeman, 1977.

goldfish across the entire tectum have been investigated by Schmidt and his
colleagues. They found (Schmidt *et al.*, 1977) that initially, following a half-retinal
ablation and nerve crush, an appropriate half-map was formed on the corresponding
half of the tectum by 36 days after operation. Between 140 and 180 days after
operation, expanded maps were found but the fibres going to the denervated half
of the tectum all came from a narrow strip of retina close to the edge of the lesion.
After 190 days or so the expanded projection was found to come more evenly from
the whole of the half-retina. Thus the first fibres to invade the vacant tectum were
those that originally projected nearest to it. A comparable delay in the development
of compression following half-tectal removal was noted by Gaze and Sharma (1970)
and has been investigated by Yoon (1976), although a conflicting (and as yet
unresolved) report comes from Cook and Horder (1974).

Schmidt (1977) has investigated the possibility that expansion of a half-retinal
projection results from a retinal regulation. He deflected the expanded half-retinal
projection into the normal ipsilateral tectum. Within 40 days of this operation the
ipsilateral map had formed and was found always to be restricted to the appropriate
half-tectum (Fig. 2.15). Thus the half-retina, which had already been shown to give

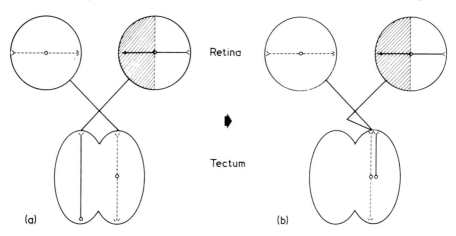

Fig. 2.15 Lack of regulation in a half-retina giving an expanded projection.
  (a) The nasal half of one retina is removed and the temporal half allowed to
expand over its contralateral tectum, then
  (b) The projection from the half-retina is induced to innervate the
ipsilateral tectum, along with fibres from the normal eye. The half-retina
now gives an appropriate half-map. After Schmidt, 1977.

an expanded projection over its own tectum, now gave a half-map when allowed
to project, along with the fibres from the normal eye, over its ipsilateral tectum. The
half-retina therefore can be said not to have regulated its specificity structure, since
it still maps over only half a normal tectum.

As a complement to this result, Schmidt (1977) suggested that, although the half-retina does not regulate, the set of tectal markers, or labels, associated with the expanded half-retinal projection, does alter; and it does so, not by a mechanism of regulation (the tectum has not been touched), but by the induction of a new set of tectal markers by the arriving abnormal retinal projection. After a half-retina had expanded its projection over its tectum, the projection from the normal eye was deflected onto this tectum. The fibres from the normal eye, instead of giving a normal map, as would be expected on a normal tectum, in all cases gave an ordered but abnormal projection: the half-retina still projected, in expanded fashion, over the tectum, and this expanded half-retinal projection was mirrored by the projection from the normal eye (Fig. 2.16). Thus a corresponding part of the normal eye gave

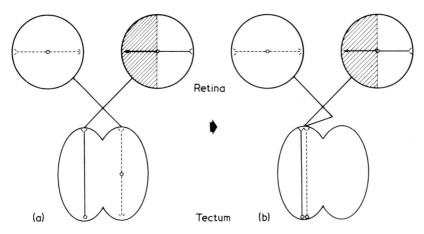

Retina

Tectum

(a)   (b)

Fig. 2.16 Alteration of tectal markers by an abnormal visual projection: 1.
　　(a) A half-retinal projection is allowed to expand over its contralateral tectum, then
　　(b) The projection from the *normal* eye is fed into this same tectum. Only the corresponding half of the normal map is now found, expanded over this tectum in a fashion similar to the projection from the half-eye.
After Schmidt, 1977.

also an expanded projection across the tectum, while the fibres from the rest of the normal eye did not appear to go to this tectum.

Thus something is certainly different about a tectum receiving an expanded half-retinal projection; and this difference is related to the abnormal projection which already covers that tectum. An even more remarkable demonstration that the tectum is somewhat altered, came from a modification of this last experiment. If the half-retinal eye was removed before the fibres from the normal eye regenerated into the experimental tectum, the normal eye was found to give a reduplicated projection over this tectum (comparable to those previously reported by Gaze and Sharma, 1970, and Sharma, 1972, in cases of field compression on a half-tectum); most electrode penetrations now gave two receptive fields, one rostral and one caudal. The rostral (nasal) series formed an expanded half-retinal projection (from the normal eye) and the caudal (temporal) series formed a more or less normal projection (Fig. 2.17).

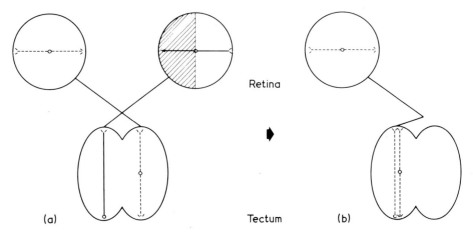

Fig. 2.17 Alteration of tectal markers by an abnormal visual projection: 2.
(a) A half-retina is allowed to expand its projection over the contralateral tectum. Then,
(b) The half-retina is removed and the normal eye is induced to innervate this same tectum. A normal eye now gives a reduplicated projection: One projection represents the normal map and the other represents an expanded map from the half of the normal eye which corresponds to the nature of the original half-retina. After Schmidt, 1977.

This experiment shows, absolutely conclusively, that new and different 'markers' now existed on the tectum; and since the only thing that had been done to that tectum was to allow a half-retinal projection to expand across it and then replace this with a projection from a normal retina, we are forced to conclude that the ordered abnormalities shown by the systems of tectal markers were put onto the tectum by the incoming optic nerve fibres themselves.

When half an eye is removed and the nerve is crushed in an otherwise normal goldfish, the optic fibres from the residual retina grow back to their original sites (Attardi and Sperry, 1963) and take some 150 days (in large fish) to develop an expanded map. Schmidt has investigated the factors underlying this delay in the formation of the expanded projection (1977). It seemed that both phenomena could be related to the persistence of fairly stable tectal cell markers, or to debris left behind by the previous projection (Murray, 1976); and it was thus of interest to see whether the changes observed could also take place in the absence of fibre/cell contacts. One tectum was therefore de-afferented for six months in several fish and after that time half of the retina of the remaining eye was removed and the fibres from the residual half-retina were deflected into the de-afferented tectum. This manoeuvre led to the immediate expansion of the half-retinal projection, without the usual intervening stage of formation of a half-map; and Schmidt (1977) argues that this is because the tectal markers associated with the previous projection had been allowed to disappear during the prolonged period of tectal de-afferentation. In the absence of tectal markers the incoming projection from the half-retina was able to establish right away an expanded projection — and induce a new and modified set of markers. Thus, if no markers remain on the tectum from the previous projection, an incoming projection can act in a 'systems-matching' fashion and fill the available space.

2.10 PRESENT STATE

The series of experiments just mentioned, in conjunction with others described in earlier sections, strongly suggest several things:

(a) Markers, or specificity labels, exist on the tectum; on tectal cells or in the form of fibre debris. These labels may be used in an affinity mechanism in the production of maps (Attardi and Sperry, 1963; Hope, Hammond and Gaze, 1976).

(b) Markers may change on the tectum as a result of different innervation (Schmidt, 1977).

(c) In the absence of optic innervation, markers eventually disappear (Schmidt, 1977). If so, then;

(d) A normally ordered and oriented map may still be set up in the absence of these markers (Schmidt, 1977) which are then induced onto the tectum by the incoming fibres. If this applies during neurogenesis as well as during regeneration then specificity markers are not required for the original setting up of the map.

If this last point is valid, we are led to ask what is the biological point of the markers in the first place. If normal projections can be established without tectal markers, and these exist mainly to assist regeneration in various experimental situations, how could we account for the evolution of such markers? We cannot make a convincing argument for an evolutionary pressure resulting from exposure of the animal to experimentally minded neurobiologists. The circumstances in which tectal markers are really useful seem never likely to occur in the normal life of the animal.

The indication that maps may be set up on the tectum without the aid of tectal specificity markers takes us back to the suggestion made earlier (and see Gaze and Hope, 1976), that time-position and fibre-sorting mechanisms may well be of great significance in neural development. If tectal markers are not necessary, then either a fibre-sorting mechanism akin to the arrow model (Hope, Hammond and Gaze, 1976) is needed, or else the fibres must arrive at the tectum in a pre-established order and properly oriented. Furthermore, if tectal markers can disappear after prolonged tectal de-afferentation, the graft translocation experiments mentioned previously (Hope, Hammond and Gaze, 1976) might give different results when performed on a long-term denervated tectum. We might then find a normal map over the grafted tectum. Such experiments are in progress.

A mechanism such as the arrow model supposes that the retinal fibres arrive at the tectum in a scrambled order; and further, that they are initially spread across the tectum in random distribution before the ordering mechanisms start to operate. Both these assumptions are likely to be invalid. The fibres in the goldfish optic nerve and pathway are known to be retinotopically ordered all the way along (Horder, 1974); and the fibres in the *Xenopus* optic tract are at least partially ordered in the diencephalon (Fig. 2.18), in that central retinal fibres lie nearest the ventricle and peripheral retinal fibres lie nearest the pial surface (Gaze and Grant, in preparation). Moreover, in the young tadpole, when the initial retinotectal connections are forming, fibres in the optic tract appear to show retinotopic order even as they approach the developing tectum, in that fibres heading for rostral tectum lie more ventrally than those heading for caudal tectum (Fig. 2.19). Retinotopic ordering of fibres in the diencephalic optic tract has also been described in *Rana* (Scalia and Fite, 1974).

Under these circumstances we can see that, if retinal fibres approach the tectum during development in an already retinotopically organized array, then the formation of a map may be facilitated to the extent that polarity information from the tectum and certain boundary conditions for the map are all that is required. While this may be the case (and it becomes increasingly likely) during development, we still have very inadequate information concerning the ordering of fibres in regeneration of the optic nerve. There is, indeed, an indication that, in goldfish at least, regenerating retinal fibres re-acquire their retinotopic order in the path beyond the lesion (Horder, 1974); but whether this is so in amphibians is not yet known.

In order to establish the role played by optic fibres in supplying markers to the tectum it is necessary to perturb the system and feed in abnormal distributions of fibres. This has been done, as we have seen, in the case of optic nerve regeneration in adult goldfish. However, for the mechanism to have general validity, it must be shown that optic fibres also specify the tectum in neurogenesis. This involves operations on eye or tectum before the first optic fibres establish tectal connections.

Only two sets of experiments up to the present time seem to meet our requirements; and these are both probably compatible with the hypothesis that optic fibres specify the tectum. The first series involves uncrossing the optic chiasma after metamorphosis in *Xenopus* with one 'compound eye' (Straznicky, Gaze and Keating, 1971). Re-examination of the published results indicates that in several cases the projection from the normal eye, when fed onto the tectum previously innervated by a compound eye, gave a map confined to the appropriate half-field. In one case (Straznicky, Gaze and Keating, 1971; Fig. 7) the situation is other than described in the paper; in this experiment both halves of the double-nasal field (TT eye) projected over the rostral two-thirds of a tectum and the nasal field of the normal eye projected simultaneously to the same region; however, the temporal

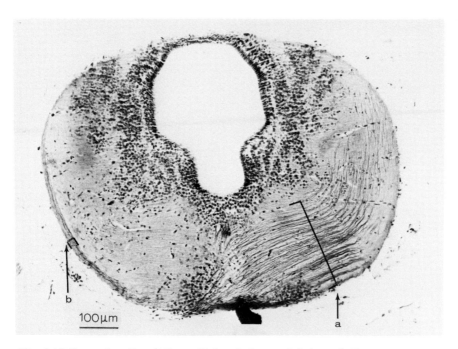

Fig. 2.18 Coronal section (15 μm; Holmes' silver stain) through the diencephalon in a *Xenopus* tadpole in which one optic nerve had been cut extracranially ten days prior to electrophysiological mapping, which showed that the retinotectal projection had recovered.

(a) The bracket shows the distribution of optic fibres on the normal side. These occupy a region from close to the ventricle to the outer margin of the diencephalon.

(b) The bracket shows the distribution of regenerated optic fibres. These are restricted to a tight band running up the far lateral edge of the diencephalon. Investigation shows that all newly arriving optic fibres run up the lateral margin of the diencephalon, at whatever stage of development they arrive. Thus the distribution of optic fibres shown on the normal side represents a retinotopic ordering of the fibres such that the 'oldest' fibres, from near the optic nerve head, run closest to the ventricle, while the 'youngest' fibres, from the present retinal margin, lie at the lateral edge of the diencephalon.
From Gaze and Grant, unpublished.

field of the normal eye projected alone to the caudal one-third of this tectum. These results, therefore, may be compatible with the idea that optic fibres specify the tectum; and that a respecification of varying degrees of completeness occurred after uncrossing the chiasma. These experiments are being extended.

The second series of experiments which offers support for the idea that optic

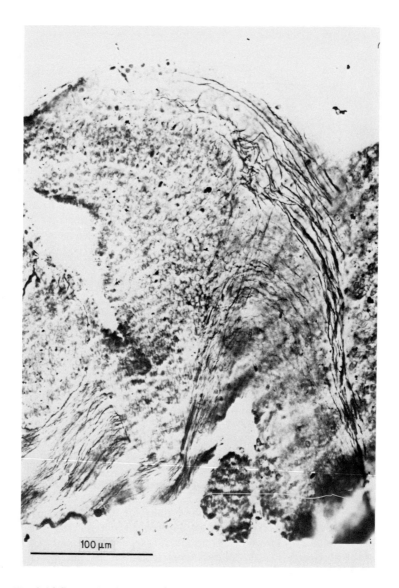

Fig. 2.19 Parasagittal section (15 μm; Holmes' silver stain) of tectum and diencephalon of a normal *Xenopus* tadpole of stage 46, showing optic nerve fibres approaching the tectum and arborizing. Dorsal is upwards, rostral to the right of the photograph. Fibres heading towards the caudal end of the tectum lie most dorsally while those arborizing at the front of the tectum lie most ventrally.

fibres specify the tectum, is that of Chung and Cooke (1975). These authors mapped the visual projection after embryonic rotation of presumptive midbrain regions in *Xenopus* and their results led them to suggest that the tectum does not acquire permanent positional labels before the arrival of optic fibres. Chung and Cooke (1975) rotated the presumptive midbrain region in *Xenopus* embryos of stages 21–24 (Nieuwkoop and Faber, 1956) and in some animals at stage 37. Later electrophysiological and histological analysis of the results showed that:

(a) The polarity of the tectal structure for map formation could be separated from its histological polarity.

(b) Map polarity seems to be associated with the relative position of certain diencephalic structures (a suggestion also made by Scalia and Fite, 1974), so that,

(c) When a diencephalic structure developed at the rostral and caudal ends of a tectum, that tectum showed two superimposed maps of opposed rostro-caudal polarity.

(d) The mapping polarity of the tectum remains modifiable by extra-tectal influences until just prior to the formation of retinotectal connections.

The only developing visual system that has been investigated extensively so far is that of *Xenopus*. Here, at early stages in the formation of the normal retinotectal projection, the fibres arriving at the tectum are already ordered in several respects (Figs. 2.18 and 2.19), and the central-peripheral retinotopic ordering in the diencephalon is a function of time of arrival of the fibres. Whether the diencephalic optic fibres going to the tectum are also ordered in relation to retinal circumferential co-ordinates is not yet established for *Xenopus*; but see Scalia and Fite (1974). It may turn out, however, that the ordering of optic fibres in the diencephalon, as they approach the tectum, is sufficient to permit the establishment of the map, with minimal additional help in the way of polarity information and information on boundary conditions. Once the map is established, long-term markers may be induced onto the tectum by the fibres. While these comments apply to amphibians and fishes, the evidence presently available suggests that the situation may be different in the chick, where early retinal lesions lead to the later appearance of 'holes' in the fibre projection to the tectum (Crossland *et al.*, 1974).

Whether or not tectal markers are used in the setting up of the map, any map mechanism except an arrow-type model or a time-position mechanism, requires that retinal markers of some sort exist. If a Sperry-type affinity system is at work, both retina and tectum must carry markers. If a fibre-sorting mechanism is at work, without tectal specificity markers, we still need retinal markers. These are needed as a basis for either fibre-sorting or for the maintenance of order in the pathway.

2.11 CONCLUSIONS

We conclude that tectal specificities, or markers, exist although they may only be used in certain circumstances. It seems very likely that the markers are put on the tectum (either induced in the tectal cells or related to fibre specificities at the sites of axonal termination) by the arriving fibres. If a set of markers already exists on the tectum (as in the normal case of rapid regeneration) when a new and different consortium of fibres arrives, a fight takes place before the eventual establishment of a new distribution of markers. The question then arises: What is the nature of specificity markers and what is the nature of affinity? Neither of these questions can yet be answered. Recent experimental reports dealing with retinotectal adhesiveness (Barbera *et al.*, 1973) and retino-retinal adhesivity (Gottleib *et al.*, 1976) are of great interest but their relevance is yet to be established. While it is customary to talk of selective adhesiveness as a main part of selective affinity, there is so far no evidence at all that adhesiveness is involved in any way whatever.

REFERENCES

Attardi, D.G. and Sperry, R.W. (1963), Preferential selection of central pathways by regenerating optic fibers. *Exp. Neurol.*, **7**, 46–64.

Barbera, A.J., Marchase, R.B. and Roth, S. (1973), Adhesive recognition and retinotectal specificity. *Proc. natn. Acad. Sci., U.S.A.*, **70**, 2482–2486.

Chung, S.H. and Cooke, J. (1975), Polarity of structure and of ordered nerve connections in the developing amphibian brain. *Nature*, **258**, 126–132.

Chung, S.H., Keating, M.J. and Bliss, T.V.P. (1974), Functional synaptic relations during the development of the retinotectal projection in amphibians. *Proc. R. Soc. Lond. Ser.* **B187**, 449–459.

Cook, J.E. and Horder, T.J. (1974), Interactions between optic fibres in their regeneration to specific sites in the goldfish tectum *J. Physiol.*, **241**, 89–90p.

Cronly-Dillon, J.R. and Glaizner, B. (1974), Specificity of regenerating optic fibres for left and right optic tecta in goldfish *Nature*, **251**, 505–507.

Crossland, W.J., Cowan, W.M. and Rogers, L.A. (1975), Studies on the development of the chick optic tectum IV. An autoradiographic study of the development of retinotectal connections. *Brain Res.*, **91**, 1–23.

Crossland, W.J., Cowan, W.M., Rogers, L.A. and Kelly, J.P. (1974), The specification of the retino-tectal projection in the chick. *J. comp. Neurol.*, **155**, 127–164.

Freeman, J.A. (1977), Effect of α-bungarotoxin on amphibian retinotectal synapses: a possible regulatory function of acetylcholine receptor in synaptic maintenance. (In preparation).

Gaze, R.M. (1970), 'The Formation of Nerve Connections', London, Academic.

Gaze, R.M. (1974), Neuronal Specificity. *Br. med. Bull.*, **30**, 116–121.

Gaze, R.M., Chung, S.H. and Keating, M.J. (1972), Development of the retinotectal projection in *Xenopus. Nature*, **236**, 133–135.

Gaze, R.M., and Hope, R.A. (1976), The formation of continuously ordered mappings. In: *Perspectives in Brain Research* (Corner, M.A. and Swaab, D.F. eds.), **45**, pp. 327–357, Elsevier, Amsterdam.

Gaze, R.M., Jacobson, M. and Székely, S. (1963), The retino-tectal projection in *Xenopus* with compound eyes. *J. Physiol.*, **165**, 484–499.

Gaze, R.M. and Keating, M.J. (1972), The visual system and 'neuronal specificity', *Nature*, **237**, 375–378.

Gaze, R.M., Keating, M.J. and Chung, S.H. (1974), The evolution of the retinotectal map during development in *Xenopus*. *Proc. R. Soc. Lond. Ser.*, **B185**, 301–330.

Gaze, R.M. and Sharma, S.C. (1970), Axial differences in the reinnervation of the goldfish optic tectum by regenerating optic nerve fibres. *Exp. Brain Res.*, **10**, 171–181.

Gottlieb, D.I., Rock, K. and Glaser, L. (1976), A gradient of adhesive specifcity in the developing avian retina. *Proc. natn. Acad. Sci. U.S.A.*, **73**, 410–414.

Hope, R.A., Hammond, B.J. and Gaze, R.M. (1976), The arrow model: retinotectal specificity and map formation in the goldfish visual system. *Proc. R. Soc. Lond. Ser.*, **B194**, 447–466.

Horder, T.J. (1971), Retention, by fish optic nerve fibres regenerating to new terminal sites in the tectum, of 'chemospecific' affinity for their original sites. *J. Physiol.*, **216**, 53–55p.

Horder, T.J. (1974), Electron microscopic evidence in goldfish that different optic nerve fibres regenerate selectively through specific routes into the tectum. *J. Physiol.*, **241**, 84–85p.

Hunt, R.K. and Frank, E.D. (1975), Neuronal locus specificity: *trans*-repolarisation of *Xenopus* embryonic retina after the time of axial specification. *Science*, **189**, 563–565.

Hunt, R.K. and Jacobson, M. (1974), Neuronal Specificity Revisited. *Current Topics in Developmental Biology*, **8**, 203–259.

Jacobson, M. (1976), Histogenesis of retina in the clawed frog with implications for the pattern of development of retinotectal connection. *Brain Res.*, **103**, 541–545.

Johns, P.A.R. (1976), Growth of the Adult Goldfish Retina. Ph. D. Thesis. University of Michigan.

Lázár, G. (1973), The development of the optic tectum in *Xenopus laevis*: a Golgi study. *J. Anat.*, **116**, 347–355.

Lopresti, V., Macagno, E.R. and Levinthal, C. (1973), Structure and development of neuronal connections in isogenic organisms; cellular interactions in the development of the optic lamina of *Daphnia*. *Proc. natn. Acad. Sci. (Wash)*, **70**, 433–437.

Meyer, R.L. and Sperry, R.W. (1973), Tests for neuroplasticity in the anuran retinotectal system. *Exp. Neurol.* **40**, 525–539.

Meyer, R.L. and Sperry, R.W. (1974), Explanatory models for neuroplasticity in retinotectal connections. In: *Plasticity and recovery of function in the central nervous system.* (Stein, D.G., Rosen, J.J. and Butters, N. eds.), pp. 45–63, Academic Press, New York.

Murray, M. (1976), Regeneration of retinal axons into the goldfish optic tectum. *J. comp. Neurol.,* **168**, 175–195.

Nieuwkoop, P.D. and Faber, J. (1956), *Normal Table of Xenopus laevis* (Daudin), North-Holland, Amsterdam.

Prestige, M.C. and Willshaw, D.J. (1975), On a role for competition in the formation of patterned neural connections. *Proc. R. Soc. Lond. Ser.,* **B190**, 77–98.

Scalia, F. and Fite, K. (1974), A retinotopic analysis of the central connections of the optic nerve in the frog. *J. comp. Neurol.,* **158**, 455–477.

Schmidt, J.T. (1977), Retinal fibers alter tectal positional markers during the expansion of the half-retinal projection in goldfish. (In preparation).

Schmidt, J.T., Cicerone, C.M. and Easter, S.S. (1977), Expansion of the half-retinal projection to the tectum in goldfish: an electro-physiological and anatomical study. (In preparation).

Scott, T.M. and Lazar, G. (1976), An investigation into the hypothesis of shifting neuronal relationships during development. *J. Anat.,* **121**, 485–496.

Sharma, S.C. (1972), Reformation of retinotectal projections after various tectal alterations in adult goldfish. *Exp. Neurol.,* **34**, 171–182.

Sharma, S.C. (1975), Visual projection in surgically created 'compound' tectum in adult goldfish. *Brain Res.,* **93**, 497–501.

Sharma, S.C. (1977), In: *Plasticity in Retinotedral Connections.* Neurosciences Research Program Bulletin. In preparation.

Sperry, R.W. (1943), Visuomotor co-ordination in the newt (*Triturus viridescens*) after regeneration of the optic nerve. *J. comp. Neurol.,* **79**, 33–55.

Sperry, R.W. (1944), Optic nerve regeneration with return of vision in anurans. *J. Neurophysiol.,* **7**, 57–70.

Sperry, R.W. (1945), Restoration of vision after crossing of optic nerves and after contralateral transplantation of eye. *J. Neurophysiol.,* **8**, 15–28.

Sperry, R.W. (1951), Mechanisms of neural maturation. In: *Handbook of Experimental Psychology.* (Stevens, S.S., ed.), pp. 236–280. Wiley, New York.

Sperry, R.W. (1963), Chemoaffinity in the orderly growth of nerve fiber patterns and connections. *Proc. natn. Acad. Sci. U.S.A.,* **50**, 703–709.

Sperry, R.W. (1965), Embryogenesis of behavioral nerve nets. In: *Organogenesis.* (De Haan, R.L. and Ursprung, H., eds.), pp. 161–186. Holt, Rinehart and Winston, New York.

Straznichy, K. and Gaze, R.M. (1971), The growth of the retina in *Xenopus laevis*: an autoradiographic study. *J. Embryol. exp. Morph.,* **26**, 67–79.

Straznicky, K. and Gaze, R.M. (1972), The development of the tectum in *Xenopus laevis*: an autoradiographic study. *J. Embryol. exp. Morph.,* **28**, 87–115.

Straznicky, K., Gaze, R.M. and Keating, M.J. (1971), The retinotectal projections after uncrossing the optic chiasma in *Xenopus* with one compound eye. *J. Embryol. exp. Morph.,* **26**, 523–542.

Willshaw, D.J. and von der Malsburg, C. (1976). How patterned neural connections can be set up by self-organization. *Proc. R. Soc. Lond. Ser.,* **B194**, 431–445.

Wolpert, L. (1971), Postional information and pattern formation. In: *Current Topics in Developmental Biology,* **6**, pp. 183–224.

Yoon, M. (1971), Reorganization of retinotectal projection following surgical operations on the optic tectum in goldfish. *Exp. Neurol.,* **33**, 395–411.

Yoon, M. (1972), Transposition of the visual projection from the nasal hemiretina onto the foreign rostral zone of the optic tectum in goldfish. *Exp. Neurol.,* **37**, 451–462.

Yoon, M. (1976), Progress of topographic regulation of the visual projection in the halved optic tectum of adult goldfish. *J. Physiol.,* **257**, 621–643.

Part 2: Experimental Analysis of Cell Positioning and Selective Cell Adhesion

3 Specific Cell Ligands and the Differential Adhesion Hypothesis: How do they fit together?

M. S. STEINBERG

Acknowledgements

The research from my laboratory described in this article was carried out with the excellent technical assistance of Mr Edward Kennedy and Mrs Doris White. It was supported by grants no. G-5779, G-10896, G-21466, GB-2315 and GB-5759X from the National Science Foundation; grants no. BC-52B and P-532 from the American Cancer Society and grant no. CA-13605 from the National Cancer Institute, DHEW.

The later development and physical testing of the differential adhesion hypothesis have been carried out jointly with Dr Herbert M. Phillips, whose knowledge and judgment have been indispensable to the progress of this work. I also thank Dr Phillips for suggesting improvements in this manuscript.

Specificity of Embryological Interactions
(*Receptors and Recognition,* Series B, Volume 4)
Edited by D.R. Garrod
Published in 1978 by Chapman and Hall, 11 New Fetter Lane, London EC4P 4EE
© Chapman and Hall

3.1 INTRODUCTION

At present there are two different formulations — the specific ligand hypothesis and the differential adhesion hypothesis — each of which purports to 'explain' spontaneous cell sorting and related phenomena. These two explanations appear to be cast at right angles to one another at best, or at worst to be altogether non-intersecting. In actuality, however, they are directed toward two different aspects of the question: 'How are the phenomena of multicellular assembly determined?' It is my purpose here to lay these relationships out as completely as I can. To do so, I must first review in some detail the differential adhesion hypothesis (DAH) and the evidence supporting it. Then it will be quite easy to see how the DAH and specific (and also not-so-specific) ligands fit into the same picture.

3.2 THE DIFFERENTIAL ADHESION HYPOTHESIS

3.2.1 Evidence from cell-sorting and tissue-spreading behaviour

The DAH grew out of the writer's efforts to distinguish whether chemotaxis ('directed movement') or 'selective adhesion' was responsible for the sorting-out of embryonic cells of different types from a cell mixture. Holtfreter (1939; 1944a,b) and his student Townes (Townes and Holtfreter, 1955) had strikingly demonstrated the self-organizing capacities of amphibian embryonic cells and struggled to explain them, ultimately concluding essentially that chemotaxis ('directed movements') drew cells and cell groups inward or outward within a tissue mass in a tissue-specific manner (specification of position), while 'cell specificity of adhesion' bound them to the preferred neighbours (specification of associations) once they were in position (Townes and Holtfreter, 1955, p. 92). I could envision how either chemotaxis by itself or intercellular adhesive differentials by themselves would be capable of specifying *both* position and associations in Townes and Holtfreter's experiments and perceived that the course of events in cell sorting would differ in certain respects depending upon which mechanism was operating. Working with reaggregates of chicken embryonic cells, I found the events of cell sorting to differ significantly from expectations based upon a chemotactic model but to agree in detail with those based upon differential cell adhesion (Steinberg, 1962b,c). I wrote (1962a):

'It is pertinent at this point to describe the essential characteristics of the cell system under discussion. (1) It is composed of discrete units of two major types. (2) The units are motile. (3) The units appear to cohere and adhere with different strengths. The same description, it will be recognized, applies to a two-phase system of mutually immiscible liquids, such as an oil-and-water system. The specific mechanisms responsible for the properties of these two radically different systems are dissimilar. Motility of cells is amoeboid, while the molecules of a liquid or the droplets in a dispersion are thermally agitated. The discreteness of cells derives from the presence of a limiting membrane. The adhesiveness of apposed cells or may not be due to the same forces which govern the mutual attraction of molecules in a liquid. But much of the most striking behavior of such two-phase liquid systems depends only upon the properties of discreteness and motility of, and differential adhesiveness among, their unitary components, and not at all upon the underlying mechanisms from which these properties are generated.'

'In view of the above comparison, one might expect many features of these cellular systems to imitate comparable features of oil-and-water systems. The very process of sorting out of two kinds of cells to produce an external tissue (continuous phase) and one or more islands of internal tissue (discontinuous phase) is a perfect imitation of the breaking of a dispersion of one liquid in another immiscible liquid of similar density. The continuous or external phase in the liquid system is that with the lower surface tension; i.e., with the lower cohesiveness among its molecules. This phase in the cell system should be that with the lower cohesiveness among its cells.'

The similarity with re-organizational behaviour in heterogeneous liquid systems suggested that other surface tension-induced phenomena familiar in liquid systems might also be displayed by cell aggregates. For example, not only will two mutually immiscible liquids demix (sort out) after being shaken together, but droplets of the two will also display characteristic mutual spreading behaviour when touched together. A droplet of that liquid which forms the continuous phase and becomes external during demixing will spread over the surface of a droplet of the liquid which forms the discontinuous phase. In both cases, the configuration stabilizes when a specific 'equilibrium configuration' is achieved. Knowledge of the 'sorting-out' behaviour of two liquids thus permits the prediction of their mutual spreading behaviour. Experiments revealed precisely the same behaviour described above for liquids in combinations of embryonic cells and tissues (Steinberg, 1962c, 1963, 1970).

As already noted, a liquid of lower surface tension will tend to spread over the surface of an immiscible liquid of higher surface tension. Hence, if one has a series of liquids, each immiscible with all of the others, their mutual spreading behaviour will follow a transitive rule, defining a hierarchy in which the liquid of highest surface tension is spread upon by — and the liquid of lowest surface tension spreads upon — all of the others. Exactly such a hierarchy was found when chick embryonic tissues of different kinds were combined in all possible binary combinations, whether by cell mixing or the fusion of intact tissue fragments (Steinberg, 1962c, 1963, 1964, 1970).

Fig. 3.1 summarizes the behavioural features described above and common to both heterogeneous multicellular and immiscible liquid systems. The striking parallels strongly suggest that the re-organizational behaviour of the multicellular system, like that of liquids, is governed by surface tensions, i.e. by *specific interfacial free energies*

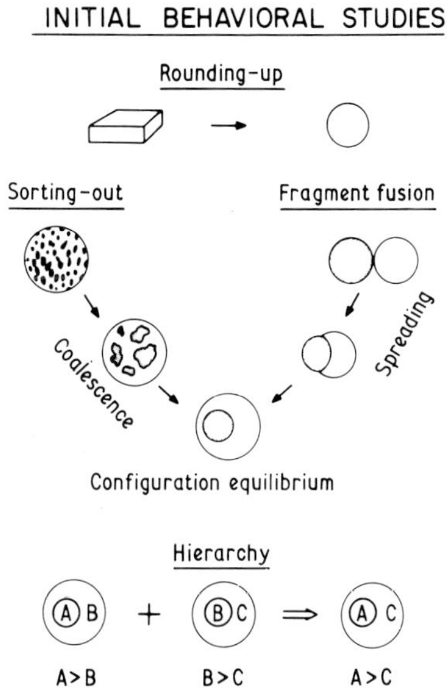

INITIAL BEHAVIORAL STUDIES

Rounding-up

Sorting-out   Fragment fusion

Configuration equilibrium

Hierarchy

A>B   B>C   A>C

Fig. 3.1 Many of the behaviours well-known in immiscible liquid systems are shown also by combinations of embryonic cells and tissues. (Reproduced from Phillips (1969) with the author's permission).

(σ's). To test this possibility it is necessary to measure the σ's of cell aggregates and to determine whether they are such as to produce the observed behaviour.

3.2.2 Specific interfacial free energies (σ's) and liquid properties of cell aggregates

The specific interfacial free energy (σ_{lo}) of a liquid l in medium o is the reversible work required to increase the area of the lo interface by a unit amount at constant temperature and volume (Phillips, 1969; Phillips and Steinberg, 1969). It can be determined, by the sessile drop method, from the equilibrium shape adopted by a droplet in a gravitational field (Fig. 3.2). The droplet's equilibrium shape is independent of its original shape; thus two qualitatively identical liquid droplets of equal volume but with different original shapes will adopt the same equilibrium shape in a given gravitational field. Only liquids show this behaviour; e.g. an elastic solid body does not. The higher a droplet's σ, the rounder will be its profile at shape equilibrium.

   The σ's of liquids are area-invariant; that is, the production of each additional unit of surface area requires the same amount of reversible work. This is in contrast to the behaviour of elastic solids, in which the production of each new unit of surface area requires more reversible work than the last. The difference arises from the fact that surface expansion of a liquid occurs by the transfer of subunits from the

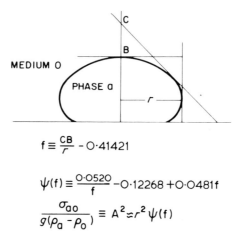

$$f \equiv \frac{CB}{r} - 0 \cdot 41421$$

$$\psi(f) \equiv \frac{0 \cdot 0520}{f} - 0 \cdot 12268 + 0 \cdot 0481 f$$

$$\frac{\sigma_{ao}}{g(\rho_a - \rho_o)} \equiv A^2 \simeq r^2 \psi(f)$$

Fig. 3.2 Geometrical construction and formulae used in the Dorsey approximation to calculate σ from the profile of a sessile drop. (Reproduced from Phillips (1969) with the author's permission).

interior to the surface, a process that is merely repeated over and over again with new interior subunits as surface expansion continues. With an elastic solid, on the other hand, the subunits remain in their original positions and stretch as the surface is expanded.

The sessile drop method would seem to be suitable for the measurement of the σ's of cell aggregates, provided that the aggregates fulfill the requirement that they reach true shape equilibrium under the conditions of measurement. This in turn requires that the change in an aggregate's surface area, in approaching shape equilibrium, occur *not* by the stretching of original surface cells (elastic solid behaviour), but by the transfer of originally internal cells into the surface (liquid behaviour). Experiments have therefore been conducted to determine whether the behaviour of cell aggregates of the kinds used in our experiments conforms to the above requirements.

3.2.3 Shape equilibrium of cell aggregates in centrifugal fields

It is common experience that embryonic tissue fragments or aggregates of many kinds round up into spherical shapes in suspension or when cultured on surfaces to which they do not adhere. This in itself means that such aggregates do tend to decrease their surface areas — and therefore their surface free energies. In preliminary experiments, Herbert Phillips and I determined that centrifugal accelerations of the order of thousands of g's, maintained for many hours, are required to deform originally spherical cell aggregates to the extent required for measuring their σ's.

Thus we needed to determine whether aggregates under these accelerations would remain healthy and assume equilibrium shapes.

From cell suspensions of chick embryonic liver, heart ventricle and limb bud, we produced aggregates in two shapes: flat and spherical. These were centrifuged in culture medium against a hard agar substratum at a constant acceleration interchange of 2000–8000 g for 24–48 hours at 37°C. Aggregates were fixed *during* centrifugation by the injection of fixative into the spinning tubes by means of a specially designed device (Phillips, 1969; Phillips and Steinberg, 1969, 1978). Control experiments conducted with the Harvey-Loomis microscope-centrifuge, in which the shape of an aggregate can be observed during centrifugation, demonstrated that injection of fixative quickly sets the shapes possessed by aggregates during centrifugation, these shapes remaining unchanged even after the centrifuge is stopped.

In some cases fixative was not injected, but the centrifuge was simply turned off after 24 hours and the aggregates recovered and observed during subsequent.culture at 1 g. These partially flattened aggregates rounded up slowly in the course of some hours, just as freshly isolated tissue fragments do; and heart aggregates and fragments continued to beat after removal from the centrifuge. Together with the normal histological appearance of centrifuged aggregates (Phillips, 1969; Phillips *et al.*, 1977a), these observations all indicated their healthy condition after centrifugation.

When initially flat and initially round aggregates were fixed as described above after 24 or 48 hours of centrifugation, it was found that the initially round aggregates had flattened down, and the initially flat aggregates had rounded up, to achieve similar, intermediate shapes (Fig. 3.3). Moreover, these shapes differed significantly

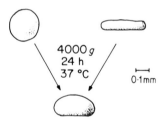

Fig. 3.3 Initially round and initially flat aggregates of chick embryonic heart cells, cultured for 24 h in a centrifuge at 4000 g, adopt the same equilibrium shape, shown here in profile. To achieve this shape, the initially flat aggregates must round up against the centrifugal force.

from tissue-type to tissue-type. Thus, the requirement was satisfied that aggregates reach shape equilibrium in a gravitational field strong enough to deform them significantly.

The flattening of an initially spherical aggregate under centrifugation might be explained as due to elastic or plastic deformation under stress from the applied force. But the rounding-up of an initially flat aggregate *against* the applied force cannot be

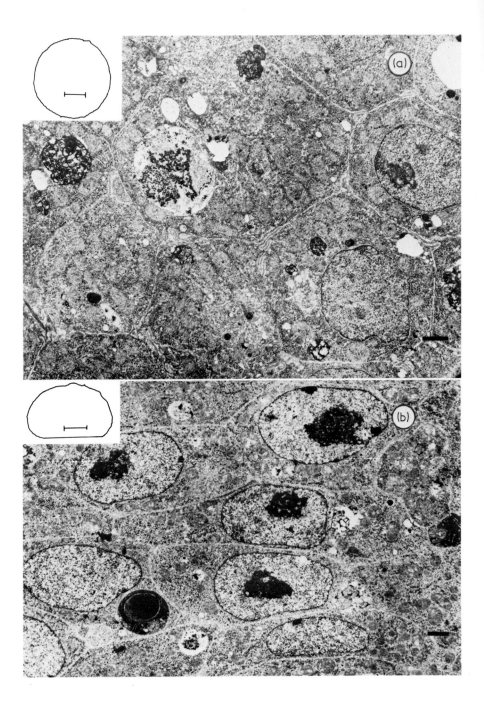

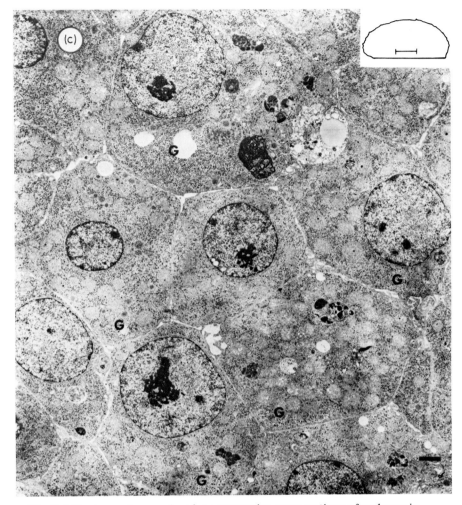

Fig. 3.4 Electron micrographs of representative cross-sections of embryonic chick liver cell aggregates fixed (a) before centrifugation, (b) starting after the fifth minute of 2000 g centrifugation and (c) during the thirty-sixth hour of 2000 g centrifugation. G indicates regions of densely packed glycogen particles. Bars represent 1 μm lengths. Inserts: tracings of vertical profiles of whole aggregates in thick sections cut adjacent to the thin sections shown in the electron micrographs. Bars represent 50 μm lengths. (Reproduced from Phillips *et al.*, (1977a), with permission of Academic Press, Inc.).

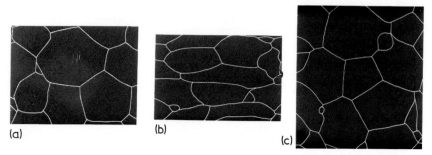

Fig. 3.5 Cell perimeters traced from the electron micrographic cross-sections in Fig. 3.4a–c. (Reproduced from Phillips *et al.* (1977a) with permission of Academic Press, Inc.).

explained in this way. Rather, it would appear to result from internalization of originally external cells due to the formation of adhesions between them, just as is the case when irregular aggregates round up at 1 *g*. As in ordinary liquids, then, the centrifugal force tending to flatten the aggregate must eventually be balanced by the surface tension (arising from the intercellular cohesive forces) tending to make it round up. As has already been indicated, this implies that a mound-shaped aggregate at shape equilibrium under 2000 *g* has a greater surface area than a spherical aggregate at shape equilibrium under 1 *g not* because its surface cells have been stretched, but because originally interior cells have been added to its surface. Nor should *interior* cells be stretched in centrifugally flattened aggregates at shape equilibrium, for, by definition (Symon, 1971), the interior of a liquid body at shape equilibrium must be free of stress.

These predictions have been tested by fixing initially spherical liver aggregates before centrifugation and five minutes or thirty-six hours after the initiation of 2000 *g* centrifugation and comparing both the aggregate profiles and the shapes of the constituent cells as seen in light-and electron micrographs of thin sections (Phillips *et al.,* 1977a). The internal cells of uncentrifuged liver aggregates were seen to be cuboidal (Figs. 3.4a and 3.5a). Surface cells varied in shape, some being cuboidal, others being somewhat columnar or flask-shaped. When the centrifuge was turned on, the aggregates promptly responded by flattening noticeably (see also Phillips and Steinberg, 1978). Both their surface and interior cells were correspondingly stretched (Figs. 3.4b and 3.5b), revealing patterns of stress within the aggregates. After thirty-six hours of centrifugation, aggregates were considerably flatter than at five minutes; but their component cells, instead of becoming more stretched than before, had returned to their original shapes (Figs. 3.4c and 3.5c), indistinguishable from those of cells within uncentrifuged aggregates. Light micrographs of liver aggregates fixed in the 25th hour of 6000 *g* centrifugation clearly showed both the relaxed configurations of internal cells and the fact that these

Fig. 3.6 Light micrograph of cross-section of embryonic chick liver aggregate fixed during the twenty-fifth hour of 6000 *g* centrifugation. Due to their movements, the cells have regained their relaxed configurations, and the originally spherical aggregate is now many cells wider than it is tall. (After Phillips *et al.*, 1977a).

initially spherical aggregates were now many cells wider than they were tall (Fig. 3.6). Thus, the conditions required for the measurement of aggregate σ's by the aggregate centrifugation method have been satisfied, with the demonstrations that centrifuged aggregates do reach shape equilibrium, and that they do so through cell rearrangements that relax transient internal stresses. The detailed evidence characterizing these embryonic cell aggregates mechanically as *elasticoviscous liquids* is presented in Phillips and Steinberg (1978) and Phillips *et al.* (1977a).

3.2.4 Correlation of aggregate equilibrium shapes with sorting-out and spreading behaviour

Pausing for a moment to review, the important points before us are as follows.

(1) σ's are the parameters that determine the miscibility or immiscibility and the spreading behaviour of liquids.
(2) Any body that assumes an equilibrium sessile drop shape in a gravitational field is characterized as a liquid and has a σ that is a determinant of that shape.
(3) Living aggregates of chick embryonic cells slowly adopt equilibrium shapes in a gravitational (centrifugal) field.

Can σ values for the aggregates in question be measured; then, and if so, do these correlate with the observed sorting-out and spreading behaviour?

   The σ of a sessile drop at shape equilibrium in a gravitational field can be determined from measurements of its geometry, its buoyant density and the

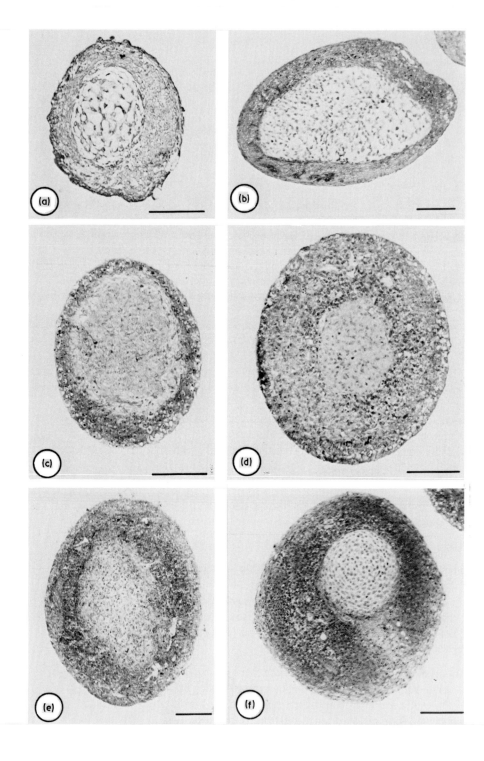

gravitational force acting upon it. Our determination of σ's from observed equilibrium shapes of centrifuged aggregates has been complicated by the development of a density gradient in the culture medium during the 24 to 48 hours of centrifugation required for the aggregates to reach shape equilibrium. Thus we are not yet in a position to calculate absolute aggregate σ's. However, measurements of the densities of the kinds of aggregates we have studied, and of the density gradients in which they have achieved shape equilibrium (Phillips, 1969 and unpublished) show that with aggregates of equal volume, their relative flatness at shape equilibrium provides a quite accurate measure of the aggregates' relative σ's. The evidence at hand is thus sufficient for a preliminary test of the ability to predict aggregate behaviour from measurements of σ or conversely; for, according to the DAH, the tendency of one mobile cell population to sort out externally to or envelop another one is determined only by their relative σ's. A cell population of lower σ will sort out externally to or envelop an (immiscible) cell population of higher σ to which it adheres.

Fig. 3.7 illustrates the inside/outside positioning behaviour of chick embryonic limb bud mesoderm, heart ventricle and liver cell populations relative to one another. Results of cell-sorting experiments are shown on the left and the corresponding tissue-spreading experiments are shown on the right. In each case, limb bud mesoderm (pre-cartilage) internalizes with respect to both heart ventricle and liver, and heart ventricle internalizes with respect to liver (Steinberg, 1963, 1964, 1970). According to the DAH, then, limb bud mesoderm should have the highest σ and liver the lowest. Equilibrium shapes of aggregates of these same tissues were determined at 4000 and 8000 g (Phillips, 1969; Phillips and Steinberg, 1969), average representatives of each group being shown in Fig. 3.8. The equilibrium shapes of the limb bud aggregates were markedly rounder on average than those of heart ventricle aggregates, and the latter, in turn, were somewhat rounder on average than those of liver aggregates, in agreement with the prediction.

The likelihood that the above hierarchies of envelopment tendency and roundness at shape equilibrium concur fortuitously is 1/6 or 0.17. Thus additional samples of this correspondence must be obtained if it is to be deemed more than co-incidental. These examples have recently been obtained.

Fig. 3.7 Equilibrium configuration adopted by chick embryonic cell and tissue combinations through the sorting out of intermixed cells (a,c,e) and the spreading of apposed, intact tissue fragments (b,d,f). Heart ventricle totally envelops limb bud pre-cartilage (a,b); liver totally envelops heart ventricle (c,d); and liver totally envelops limb bud pre-cartilage (e,f); illustrating the transitivity of inside-*vs*-outside positioning. According to the differential adhesion hypothesis, a cell population of lesser cohesiveness (σ) should tend to envelop one of higher σ. Thus the σ's of the three tissues represented here should decline in the sequence $\sigma_{LB} > \sigma_{HV} > \sigma_{L}$. Bars represent 100 μm lengths. (Reproduced from Steinberg (1964) with permission of Academic Press, Inc.).

SHAPES OF CENTRIFUGED AGGREGATES

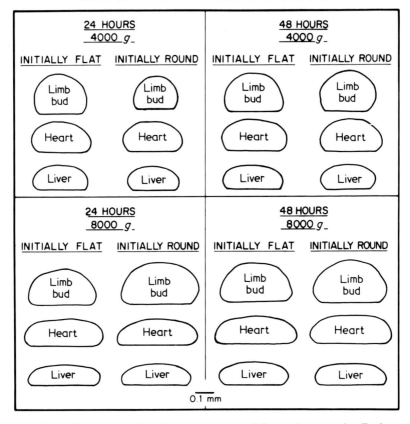

Fig. 3.8 Profiles of centrifuged aggregates traced from photographs. Each pair was derived from the same kind of primordium. The left-hand member of each pair was initially flat, while the right-hand member was initially round.

Upper left: three pairs of aggregates spun at 4000 *g* for 24 h.
Upper right: aggregates spun at 4000 *g* for 48 h.
Lower left: aggregates spun at 8000 *g* for 24 h.
Lower right: aggregates spun at 8000 *g* for 48 h.

At the end of each run, the aggregates were fixed in the centrifuge before the speed of the rotor was reduced.

We discovered that the mutual envelopment tendencies of both heart ventricle and liver aggregates could be altered experimentally (Wiseman *et al.*, 1972). For example, when undissociated fragments of 5-day chick embryonic heart ventricle were precultured in shaker flasks for 1/2 day and then fused with fragments of 5-day chick embryonic liver precultured for 2½ days, the heart fragments tended to envelop the liver fragments. However, when the heart fragments were precultured

Dissociation effects   Time–in–culture effects

(a) Behavior with liver

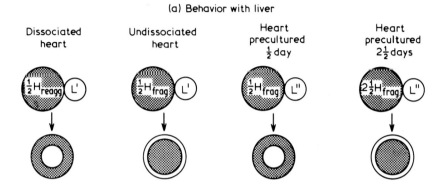

(b) Predicted envelopment behavior

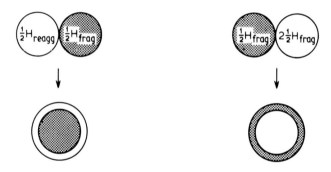

Fig. 3.9 (a) Envelopment behaviour of heart and liver cell populations reveals that dissociation and reaggregation prior to 1/2-day of preculture makes heart cell populations more externalizing ($\frac{1}{2}H_{reagg}$ vs. $\frac{1}{2}H_{frag}$), whereas longer preculturing makes heart cell fragments more internalizing ($2\frac{1}{2}H_{frag}$ vs. $\frac{1}{2}H_{frag}$). From these observations of the behaviour of heart *vis-a-vis* liver cell populations we predict, following the principle of transitivity in envelopment behaviour, that heart cell populations pretreated in these ways will envelop one another in the directions shown in (b).

for 2½ days instead of only 1/2 day, the liver fragments tended to envelop them. Interpreted according to the DAH, the cohesiveness (σ) of the heart fragments had increased during the additional 2 days of culture. When the liver masses fused with 1/2-day precultured heart fragments were reaggregates precultured for 1/2 day (rather than fragments precultured for 2½ days), they tended to envelop the heart fragments. But when the heart fragments were dissociated and reaggregated

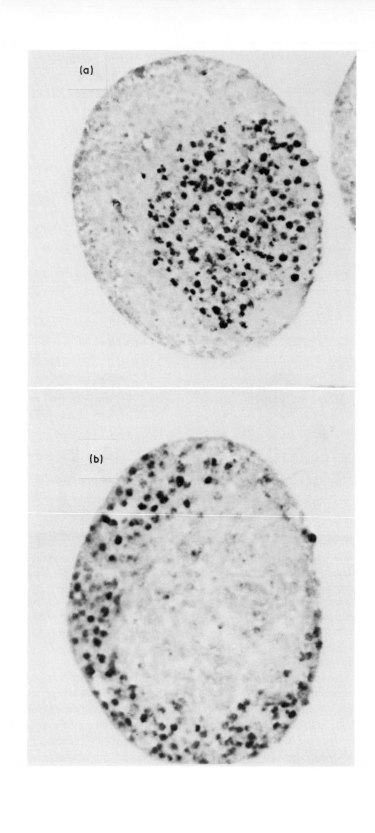

preceding their 1/2 day of preculturing, the resulting heart reaggregates tended to envelop the liver fragments. Interpreted according to the DAH, the cohesiveness (σ) of the heart fragments had decreased as a consequence of their dissociation and reaggregation. These facts are summarized in Fig. 3.9a.

The DAH makes the following three sets of predictions based upon the above observations. First, it predicts that, envelopment tendencies being transitive, undissociated heart fragments that have been precultured for only 1/2 day ($\frac{1}{2}H_{frag}$) should tend to envelop otherwise identical heart fragments that have been precultured for 2½ days ($2\frac{1}{2}H_{frag}$). Second, and for the same reason, it predicts that heart masses that have been dissociated and reaggregated preceding 1/2 day of preculture ($\frac{1}{2}H_{reagg}$) should tend to envelop otherwise identical heart masses that have remained as intact fragments the entire time ($\frac{1}{2}H_{frag}$). These predictions, both of which hold that an experimental manipulation that reverses a tissue's envelopment behaviour with a second tissue should *induce* in it, *de novo*, corresponding envelopment behaviour towards its former 'self' (provided that the two masses do not preferentially intermix), are illustrated in Fig. 3.9b. Both predictions have been confirmed (Phillips *et al.*, 1977b), as is shown in Fig. 3.10. Finally, the DAH attributes these changes in envelopment tendency to underlying changes in aggregate cohesiveness (σ), and predicts that

$$\sigma_{2\frac{1}{2}H_{frag}} > \sigma_{\frac{1}{2}H_{frag}} > \sigma_{\frac{1}{2}H_{reagg}}.$$

The buoyant densities of all three types of heart aggregate were found to be identical (Phillips, in preparation), so that the roundness of the aggregates at shape equilibrium in a given centrifugal field should accurately reflect their σ's. Masses of all three kinds were centrifuged at $1000\,g$ for 24 or 44 hours. In the case of the reaggregates, some were initially flat while others were initially round. As can be seen in Fig. 3.11, which shows for each group the profiles of masses of average roundness, the *reaggregates* had not quite reached shape equilibrium at 24 hours, but had reached it in 44 hours. Both kinds of *fragments* had evidently reached shape equilibrium by 24 hours, since their shapes did not change further during an additional 20 hours of centrifugation. Comparison of the equilibrium shapes shows that the heart fragments precultured for 2½ days before centrifugation had markedly rounder profiles than those precultured for only ½ day, while the profiles of those masses that were dissociated and reaggregated before being precultured for ½ day were markedly flatter. For a second time, then,

Fig. 3.10 Autoradiography of complete-envelopment configurations of heart aggregates cultured 48 h in centrifugation medium after fusion. (a) An unlabeled trypsin-dissociated heart reaggregate cultured ½ day prior to fusion ($\frac{1}{2}H_{reagg}$) has enveloped a [^{3}H]-thymidine-labeled (undissociated) heart fragment also precultured ½ day ($\frac{1}{2}H_{frag}$). (b) A [^{3}H]-thymidine-labeled heart fragment precultured ½ day ($\frac{1}{2}H_{frag}$) has enveloped an unlabeled heart fragment cultured 2½ days prior to fusion ($2\frac{1}{2}H_{frag}$). (Reproduced from Phillips *et al.* (1977b), with permission of Academic Press, Inc.).

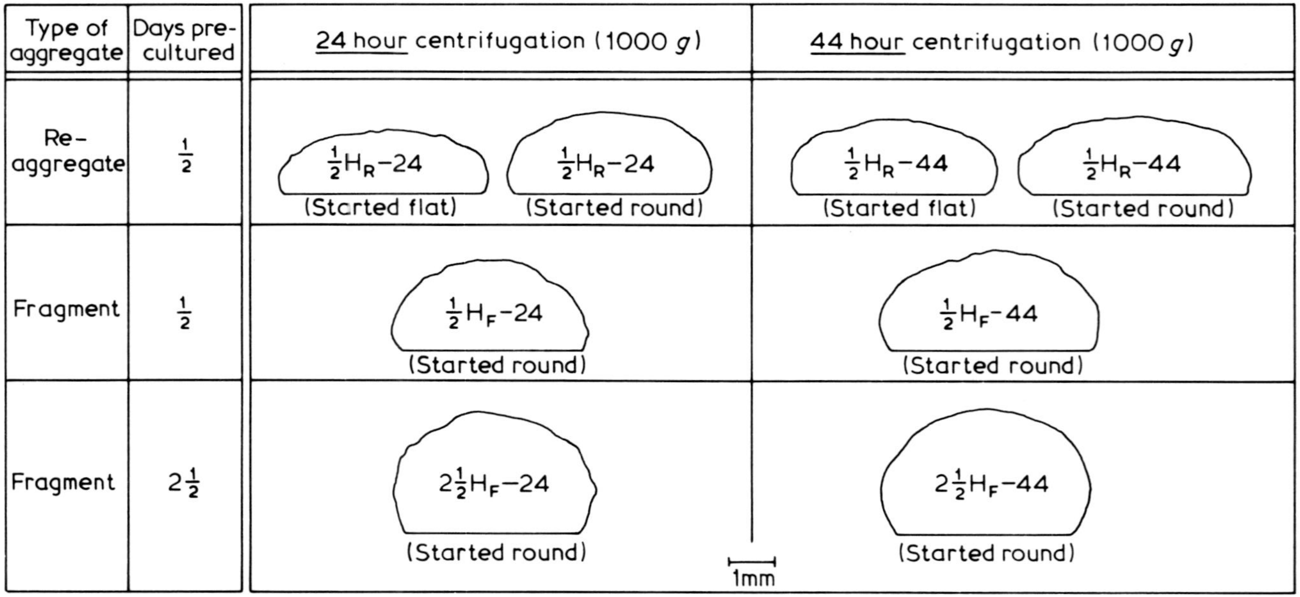

Fig. 3.11 Vertical profiles, traced from side-view photographs, of the various types of heart aggregates, centrifuged at 37°C and 1000 g for 24 or 44 h. Fixative was injected during centrifugation, 15–30 min before the rotor was stopped. (Reproduced from Phillips et al., (1977b), with permission of Academic Press, Inc.).

a perfect correspondence has emerged between envelopment tendency and aggregate roundness at shape equilibrium for a series of three 'tissues'. The likelihood that *both* sets of correspondences are merely fortuitous is $(1/6)^2 = 1/36$, or 0.03.

The above correlations between cell population behaviour and measurements of relative σ's indicate that tissue σ's very likely do determine whether two cell populations will preferentially mix or sort out and, if the latter, which of the two will tend to internalize and which will tend to spread over the other. This conclusion is justified by the observed correlations although the sessile drop (aggregate centrifugation) method measures only the σ's at liquid–medium interfaces (σ_{ao}; σ_{bo}) and not that at the interface between two liquids (σ_{ab}). However, the latter can be specified relative to the former, within certain limits, from knowledge of the geometry of the equilibrium configuration arrived at by a combined pair of liquid bodies (Steinberg, 1963, 1964). This is more easily grasped if, instead of expressing these adhesive parameters as σ's, one expresses them as W's, the closely related 'reversible works of adhesion.' This is because W_{ab}, the reversible work of adhesion of liquid a to liquid b, is fully analogous to W_{aa} and W_{bb}, the reversible works of *co*hesion of liquids a and b, respectively; is defined independently of the latter; and corresponds to one's intuitive impression of adhesive intensity. σ_{ab}, on the other hand, is defined in such a way as to be a function not only of the intensities of a–b adhesions but of the intensities of a–a and b–b adhesions as well. The relationships between σ's and W's are given in equations 3.1–3.3 (Dupré, 1869).

$$\sigma_{ao} = \frac{W_{aa}}{2}; \ W_{aa} = 2\,\sigma_{ao} \tag{3.1}$$

$$\sigma_{bo} = \frac{W_{bb}}{2}; \ W_{bb} = 2\,\sigma_{bo} \tag{3.2}$$

and,

$$\sigma_{ab} = \frac{W_{aa} + W_{bb}}{2} - W_{ab}; \tag{3.3}$$

$$W_{ab} = \sigma_{ao} + \sigma_{bo} - \sigma_{ab}.$$

The equilibrium configuration adopted by a two-phase liquid system is determined by the free energies of association (σ's; W's) among and between its components. Two liquids will be miscible if $W_{ab} \geq \frac{1}{2}(W_{aa} + W_{bb})$ and immiscible if $W_{ab} \leq \frac{1}{2}(W_{aa} + W_{bb})$. In the latter case, the liquid of lower σ will tend to spread over the liquid of higher σ, the extent of spreading depending upon the ratio W_{ab}/W_{bb}, where b designates the liquid of lower σ. When $W_{ab} < W_{bb}$, phase b will be spread only partially over the surface of phase a at shape equilibrium, while when $W_{ab} > W_{bb}$, phase b will completely envelop phase a at shape equilibrium. This can be derived as follows from equation 3.1–3.3 above and equation 3.4 (Phillips, 1969, p. 199), in which the θ's refer to the angles defined in Fig. 3.12.

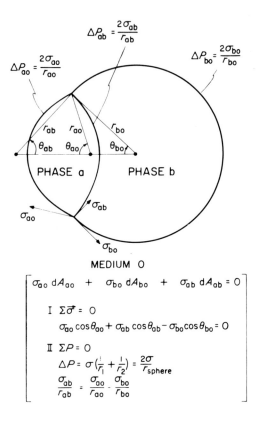

Fig. 3.12 Continuous phases a and b, with spherical interfaces, in medium o. At equilibrium, forces of magnitudes σ_{ao}, σ_{bo} and σ_{ab}, tangent to the a–o, b–o and a–b interfaces, respectively, will balance one another. From this balance of forces arise the adhesive relations and corresponding equilibrium configurations used in the differential adhesion hypothesis. (Reproduced from Phillips (1969) with permission of the author).

$$\sigma_{ab} \cos \theta_{ab} + \sigma_{ao} \cos \theta_{ao} - \sigma_{bo} \cos \theta_{bo} = 0 \qquad (3.4)$$

When phase b has just enveloped phase a, the following values obtain:

$$\theta_{ao} = 0°; \cos \theta_{ao} = 1$$

$$\theta_{bo} = 0°; \cos \theta_{bo} = 1$$

$$\theta_{ab} = 180°; \cos \theta_{ab} = -1$$

Substituting these values into equation (3.4), we obtain $\sigma_{ab} = \sigma_{ao} - \sigma_{bo}$; which through substitution of equations (3.1–3.3) becomes $W_{bb} = W_{ab}$. In other words,

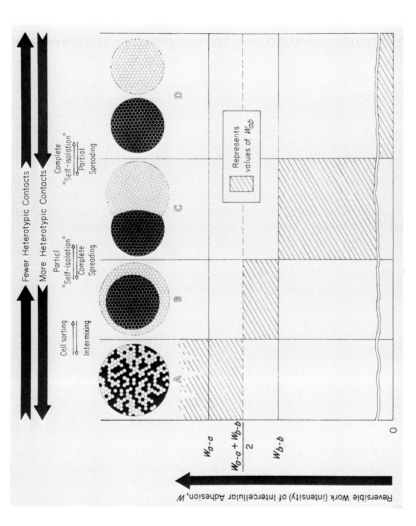

Fig. 3.13 Illustration of how the reversible works of cohesion (W_{a-a} and W_{b-b}) and adhesion (W_{a-b}) determine the most stable configuration of a liquid system. These relationships should apply to any multi-subunit system that adopts liquid-like equilibrium shapes, whether the subunits are molecules or cells. See text for full description.

when the less cohesive of two mutually immiscible liquids adheres to its partner more weakly than to itself, it will spread only partially over its partner at shape equilibrium; while when it adheres to its partner more strongly than to itself, it will envelop its partner completely. The precise relationships between the works of cohesion and adhesion in a two-phase liquid system and the equilibrium configurations they determine are illustrated in Fig. 3.13.

3.2.5 Lack of correlation of adhesion rates with sorting-out and spreading behaviour

The first attempt to measure 'intercellular adhesiveness' and relate the results to the DAH was that of Roth and Weston (1967). They measured the rate at which pre-formed chick embryonic liver and retina aggregates collected liver and retina cells from a suspension and found that aggregates of both kinds collected like cells faster than cells of the other kind. Similar results were obtained with other cell and tissue combinations also (Roth, 1968). Subsequently, many variations of the 'cell collection' procedure have been introduced (Roth *et al.*, 1971; Walther *et al.*, 1973; Barbera *et al.*, 1973; Barbera, 1975; Marchase *et al.*, 1975; McClay and Baker, 1975; McGuire and Burdick, 1976; Cassiman and Bernfield, 1976; Gottlieb *et al.*, 1976; Vosbeck and Roth, 1976). In many (but not all) cases in which tissue selectivity has been tested, it has been found that cells initiate adhesions to the tissue of origin at a greater rate than to other tissues.

Roth and Weston (1967) and Roth (1968) suggested that their measurements might cast doubt upon the DAH. Their discussion of this question contained several misconceptions that I have discussed elsewhere (Steinberg, 1970, pp. 426–428). The central issue at this point, however, is whether the free energies (σ's or W's) invoked by the DAH can be assessed by measurements of rates of cellular attachment to other cells and tissues. There are two reasons why they cannot.

First and foremost, the rate at which a reaction proceeds is not in general a measure of the change in free energy that accompanies the reaction. Many reactions with a small ΔF proceed rapidly, while other reactions with a large ΔF do not proceed at all unless primed, e.g. by high temperature. For example, oil readily spreads on water, but an explosive charge may lie dormant indefinitely. Also, catalysts greatly increase reaction rates without, of course, changing the free energy of the reaction. Such differences in reaction rates are due to differences in activation energies.

Second, because collisions between cells and other bodies in a stirred suspension are transient, adhesions must be made very quickly if they are to be made at all. Thus, only those adhesions that can be made almost instantly upon collision can contribute to the rates of adhesion measured by such collection assays. Structural junctions such as desmosomes and adherens-type junctions are not assembled this quickly upon contact (e.g. Overton, 1962, 1977; Heaysman and Pegrum, 1973), but may make a major contribution to the intensities of adhesion between cells that have been in contact for a prolonged period, as is the case with cells within populations approaching shape equilibrium. Indeed, Umbreit and Roseman (1975) have shown that the initial adhesions between aggregating embryonic cells are made in the absence of metabolic energy production and are readily broken by gentle shearing, but that these weak adhesions are succeeded in time by bonds whose appearance requires the production of metabolic energy and which are not readily separated by shearing. That the adhesion energies responsible for initial adhesion when cells collide may be a miniscule fraction of the adhesion energies that develop with prolonged contact is suggested by a comparison of measurements by Curtis and by Phillips. Curtis obtained value on the order of only $10^{-6}-10^{-5}$ erg cm^{-2} for the energy of *initial* adhesion between chick embryonic liver, limb bud and heart ventricle cells (1969), while Phillips obtained preliminary values of the order of $6-20$ erg cm^{-2} (a difference of $6-7$ orders of magnitude!) for the σ's of aggregates of the same tissues at shape equilibrium after more than a day in culture (Phillips, 1969).

In order to test directly whether measurements of adhesion *rates* could measure the *intensities* of adhesion that in turn can explain cell sorting and tissue spreading, an experimental system was designed in which, for the first time, a–to–a, b–to–b and a–to–b adhesion rates could be measured between pairs of preformed aggregates. Such rates (R's) were measured and showed the same 'preference' for self-adhesion reported earlier by Roth using a cell collection method

$$(R_{h-h} > R_{r-r} > R_{h-r} \text{ and } R_{l-l} > R_{r-r} > R_{l-r}).$$

Adhering aggregate pairs were then cultured further, as were corresponding mixed aggregates (sorting-out) and their equilibrium configurations were determined. These were then compared with the theoretical equilibrium configurations calculated on the hypothesis that the measured adhesion *rates* were proportional to the corresponding W's at shape equilibrium. In the 'theoretical' configuration, neural retina covered either liver or heart ventricle only partially, whereas in actuality it enveloped both liver and heart ventricle completely (Moyer and Steinberg, 1976), corresponding with the sequences

$$W_{h-h} > W_{h-r} > W_{r-r} \text{ and } W_{l-l} > W_{l-r} > W_{r-r}.$$

Other compelling evidence that rates of cell adhesion can vary independently of the determinants of cell sorting and tissue spreading comes from experiments on the effects of embryonic age on these cell properties. The aggregation rate of chick embryonic neural retina cells decreases dramatically and steadily between the 5th and the 19th day of development (Moscona, 1961, 1962; Gershman, 1970). Yet older and younger retinal cells, despite great differences in their aggregation rates, do not sort out from one another when mixed together; nor do fragments of undissociated retinal tissue of different ages preferentially envelop one another after they are experimentally fused (Gershman, 1970).

It is sometimes stated that progress in the understanding of cell adhesiveness and its roles in morphogenesis has been hindered by a lack of quantitative measurement procedures, and that the development of the cell collection assays has remedied this lack. As we have seen, however, these kinetic assays, useful as they are for studies of certain kinds, do not measure the adhesive parameters that govern the multicellular assembly processes with which we have been concerned. In fact, there can exist no single parameter that can be singled out from all others and exclusively identified as 'adhesiveness'. Unlike such uniquely defined entities as *mass, velocity, force* and the like, 'adhesiveness' is an informal word that can be applied to many phenomena associated with attraction between molecules without uniquely characterizing any of them. Thus, friction, tensile strength, ductility, viscosity, surface tension and vapour pressure are all adhesion-related phenomena, but are nevertheless separate and distinct entities, each of which must be measured in its own way. It is an illusion to suppose that any single, arbitrarily chosen assay procedure will necessarily measure all aspects of cell 'adhesiveness' that one might wish to measure. Rather, the design or selection of a procedure for measurement should have a sound theoretical basis, arising from a rigorous analysis of the physics of the system to be studied.

3.2.6 What underlies the σ's of cell aggregates?

The proposal that σ's, the specific interfacial free energies of cell aggregates, are the determinants of cell sorting and tissue spreading behaviour has been supported by evidence reviewed earlier in this article, including assays for σ. What can be said about the microscopic properties undelrying aggregate σ's? Can σ's be explained as purely and simply the consequence of intercellular adhesions?

As state earlier, σ_{lo} of a liquid body l in medium o is the reversible work required to increase the area of the l–o interface by a unit amount at constant temperature and volume (Phillips, 1969; Phillips and Steinberg, 1969). It is important to realize that a liquid is by definition a body having certain mechanical properties (Symon, 1971), and that cell aggregates capable of sorting out and spreading behaviour have been shown to possess these properties, as described above. The definition of σ, the criteria for its action and the operations by which it is measured are all rooted in thermodynamics. Every liquid surface has a σ which is measured by determining the *reversible* work (W) needed to expand the surface by a unit amount.[1] The σ of a liquid body will reflect the sum of the W's *of all processes required to expand its surface.* The most obvious such process is the transfer of originally internal subunits into the surface, the most obvious W being that required to overcome the bonding of the new surface-to-be-exposed to its original connections.[2] But there is another potential component of σ, in both ordinary liquids and cell aggregates. When a subunit is moved from the interior into the surface of a liquid body, it becomes exposed to anisotropic forces which may change its shape. This shape-change may or may not have a W associated with it; but if it does, this W will be included in σ. In the case of a cell aggregate, then, if the surface cells differ in shape from those in the interior, σ will reflect not only the W required to overcome intercellular bonding but also whatever W might be required to produce the shape change of cells as they move from internal positions into external ones (Fig. 3.14). It should be noted,

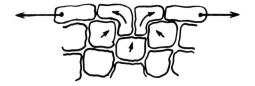

Fig. 3.14 Illustration of the manner in which the surface area of a liquid cell aggregate under tension is increased. Cells may at first be stretched, but originally subsurface cells continue to come to the aggregate's surface until the aggregate is again at shape equilibrium with the forces acting upon it. Because the surface of a liquid is exposed to anisotropic forces, surface cells may have a different shape than subsurface cells.

however, that the existence of a shape difference between external and internal cells of an aggregate does not in itself imply that reversible work is necessary to produce it. Particularly if the adhesive areas of a cell are localized and it has excess membrane, it may exist in many thermodynamically equivalent configurations (see also[2]). It is also important to note that the transient stretchings of internal cells that we have observed in centrifuged aggregates *before* shape equilibrium is reached are no longer present when shape equilibrium has been achieved (Figs. 3.4–3.6; Phillips *et al.* 1977a). Thus, the work required to produce these transient stretchings of cells, although real, does not enter into aggregate σ's.

Although we have as yet no way to apportion an aggregate's σ between different compartments of reversible work, it seems likely that the major part of σ consists of reversible work required to overcome the bonding of cells to one another and to extracellular structures as cells are brought from an aggregate's interior into its surface. Thus σ must reflect the properties of whatever it is that holds cells together.

3.2.7 Cell ligands and adhesive specificity

What *does* hold cells together? At the ultrastructural level, cells are seen to adhere directly to each other by means of a variety of contact specializations such as gap junctions (nexuses), tight (occludens) and intermediate (adherens) junctions or desmosomes, and by unspecialized regions of apposed membrane. In addition, many cells adhere to extracellular structures such as collagen fibers and glycosaminoglycans or proteoglycans (reviewed in Wessells, 1977). Bringing areas of cell surface having these associations out onto the surface of an aggregate requires the breaking of these associations, and the W's required to achieve this must inevitably enter into the corresponding σ's.

Little is yet known about the chemistry of bonding among embryonic cells or between such cells and extracellular structures. However, much attention has been given to the hypothesis that the tissue-specific sorting out of cells is a straight-forward reflection of tissue-specific adhesive 'factors' on their surfaces (Moscona, 1962, 1963, 1968; Moscona and Moscona, 1962; Lilien and Moscona, 1967). The simplicity and directness of this formula − that specificity is neither gained nor lost but carried straight along from molecule to cell to cell population − together with some evidence for tissue-specific promotors of cell aggregation, have made it an appealing one; and it has been presented (Moscona, 1968, p. 273)[3] as being inconsistent with the differential adhesion hypothesis. These are matters that I now hope to clarify.

What is the evidence for embryonic tissue- or cell-specificity of association? Since every tissue contains cells of more than one kind, it clearly cannot be that cells adhere only to others of their own kind. It has been suggested that cells' 'recognition' markers are coded at the tissue rather than the cellular level. However, even embryonic cells and tissues that never encounter one another in the embryo and have no known strategic reason for 'recognising' one another (e.g. limb bud mesoderm and retinal pigmented epithelium; Steinberg, 1963) readily adhere when placed in contact; and cells from a single tissue but of different types (e.g. retinal prospective photosensitive cells and ganglion cells) sort out after being mixed together.

At the ultrastructural level, undifferentiated close contacts and *adherens*-type junctions or desmosomes are both formed between cells of different types sorting out within mixed aggregates (Armstrong, 1970, 1971; Overton, 1977). Taken together with the constancy of structure of these junctions from tissue to tissue, this points toward an underlying similarity in their chemical composition in cells of different types. *Adherens*-type junctions or desmosomes being regions of intercellular adhesion, it seems likely that at least one type of adhesion site is shared by cells of many types. It is not known whether the apposed halves of the undifferentiated close contacts formed between heterotypic cells are also similar in their molecular composition. This does not mean, nor do I believe or intend to imply, that *all* cell surface adhesion sites within an organism are common to all of its adhesive cells. But it *does* suggest that the capacity of embryonic cells of widely different types to adhere to each other very likely reflects a sharing of certain types of adhesion site. To state that most of the adhesive cells in an embryo can adhere to one another is not, of course, to state that they do so with equal intensities.

3.3 SPECIFIC CELL LIGANDS, TISSUE ASSEMBLY AND THE DAH

I have reviewed evidence from several directions and sources which indicates that the energies of initial adhesion between aggregating cells are far weaker than those which, developing over the course of some hours, govern cell sorting and tissue spreading. I have also reviewed evidence showing that the rates of initiation of cell adhesions can vary independently of the differentials that govern cell sorting and tissue spreading. These facts mean that aggregation assays cannot be relied upon to measure the determinants of tissue assembly behaviour. Yet the adhesion-promoting effects of the 'factors' or 'cell ligands' whose specificities have been invoked to explain cell sorting have all been demonstrated in cell aggregation assays. There exists, to my knowledge, no evidence that these agents function in or have any effect upon the cell sorting behaviour they are invoked to 'explain'. I do not claim that they are *not* involved in the control of this behaviour, but wish only to under-score the fact that there is at this time no evidence one way or the other.

Moscona has presented the difference between the DAH and his specific adhesive factor hypothesis as a choice between alternatives: an explanation (he calls it a 'notion') invoking 'quantitative differences exclusively' (1968)[3] versus one invoking 'qualitative differences between specific constituents on the cell surface as being primarily responsible for selective attachment and histogenetic affinities of cells'. In doing so, he conveys the false impression that both hypotheses deal in the same terms − that the DAH represents cells as having exclusively qualitatively similar adhesion sites on their surface − despite absolutely explicit statements to the contrary, starting with the very first presentations of the DAH[4]. As I have repeatedly made clear, however, the DAH 'is not a chemical hypothesis. It deals with the *strengths,* or more accurately, the *works* or *energies* of the various cell adhesions possible within a system, and it is indifferent to the chemical or physical means by which those adhesive works or energies are generated' (Steinberg, 1970, p. 425). Since σ's are scalar quantities, meaning that one could arrange the adhesive energies or σ's of all possible cell population interfaces on a single scale from the lowest to the highest, the DAH does deal 'with quantitative differences exclusively'. But this fact has nothing whatever to do with the question of the similarities or dissimilarities between adhesive molecules on the surfaces of different kinds of cells. There is, however, a meeting place between the DAH and questions concerning the molecular specificity and cellular distribution of substances mediating the formation of inter-cellular adhesions. This lies in the consideration of the equilibrium configurations of cell populations and how they are determined (see also Steinberg, 1964).

3.3.1 How equilibrium configurations bear upon adhesive specificity

Interactions between molecules on cell surfaces, potentially occurring through a variety of mechanisms (reviewed in Curtis, 1973), may overcome the electrostatic repulsive forces between cells, causing the cells to adhere more or less strongly to one another. The more numerous the bonds that form between two cells, and the greater their strength, the more tightly will the cells adhere. Cell type- '*specificity*' of adhesion, however, arises neither from the use of any particular molecular mechanism of adhesion nor from the development of any particular intensity of adhesion between a pair of cells. It arises, rather, from *differences* in the adhesive intensities that develop when cells of a given kind are paired with one or another kind of cell. Since most if not all adhesive cells can adhere to a variety of cellular and non-cellular surfaces, although not necessarily with equal intensity, I prefer to speak of the 'selectivity' rather than the '*specificity*' of cell adhesion.

The 'specific ligand' hypothesis assumes the following straightforward relationship.

Molecular site
distribution   Cell adhesion   Tissue organization

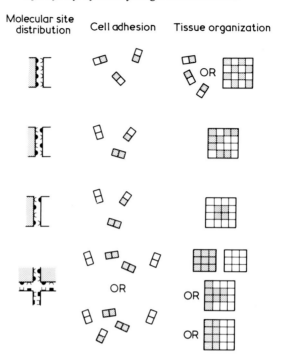

Fig. 3.15 Illustration representing how the affinities and cellular distribution of molecular binding sites on cell surfaces lead to particular sets of intercellular adhesive intensities, and how the latter in turn determine particular most stable tissue configurations. See text for details.

'Ligands' with chemical binding specificity are distributed among cells in a tissue-specific way, thereby conferring tissue specificity of adhesion upon these cells. This tissue specificity of adhesion in turn causes the cells, when combined in large numbers with cells of other types, to sort out into tissue-specific groupings. Instances of this simple transfer of specificity from molecule to cell to cell population exist, and I will begin by describing two of them; but some further examples will show the limitations of this simple-minded formulation as a comprehensive explanation of multicellular assembly phenomena.

When sponges are dissociated into cell suspensions and such suspensions from different species are mixed together, in certain species combinations the cells reaggregate species-specifically, showing little tendency to adhere to cells of the other species (reviewed in Humphreys, 1967). It seems most reasonable to suppose that this species-selectivity of adhesion is due to an inability of each species' adhesion-mediating molecules to 'recognize' those of the other species. The combining properties of these molecules should therefore be quite selective. Indeed, substances promoting the aggregation of these cells species-specifically have been isolated from these sponges and partially characterized (Moscona, 1968; Henkart *et al.*, 1973). This situation is represented in Fig. 3.15 in the upper part of the bottom row. Specificity here is transmitted directly from molecule to cell population.

In the case of the yeast, *Hansenula wingei,* two mating types exist, comparable to sexes, whose specific cross-reaction is due to the interaction of complementary sex-specific glycoproteins on their surfaces (Crandall and Brock, 1968). Again, specificity is transmitted directly from molecule to cell population, but in this case all the 'locks' are on the cells of one mating type while all the 'keys' are on cells of the other. This circumstance is represented in the first row of Fig. 3.15. This species shows massive agglutination when mating types are mixed, presumably because the interacting substances cover the cells. Restriction of such substances to single foci presumably limits gametes to pair-formation in other species.

Fig. 3.13 shows how the same events are described according to the DAH. The separate aggregation of sponge species falls into category D, corresponding to the sequence $W_{aa} > 0; W_{bb} > 0; W_{ab} \simeq 0$. The mixed aggregation of yeast gametes falls into category A, corresponding in this case to the sequence $W_{aa} \simeq 0; W_{bb} \simeq 0; W_{ab} > 0$. In these examples, both the DAH and the cell- or tissue-specific ligand hypothesis accurately describe the events. They do so also in a case in which the adhesion systems of two kinds of cells are identical (Fig. 3.15, 2nd row); here the specific ligand hypothesis would state that all cells recognise each other equally, while the DAH simply states that $W_{aa} = W_{bb} = W_{ab}$, yielding random mixing.

Consider now the case illustrated in row 3 of Fig. 3.15. The adhesion system here utilizes complementary molecules. Both kinds of cells utilize the *same* adhesion system, and so a–a, b–b, and a–b cell pairs are readily formed. The two cell populations, when mixed, co-aggregate to yield mixed aggregates. But within these mixed aggregates, the a–a, b–b and a–b adhesions are of different strengths because the b cells have fewer adhesion sites than the a cells. If the number of intercellular bonds (n) made is in each case proportional to the product of the areal concentrations of binding sites in the apposed surfaces, then the number of bonds made between an a cell and a b cell will equal the *geometric mean* of the number of bonds made between pairs of a cells and pairs of b cells, namely

$$n_{ab} = \sqrt{n_{aa} \times n_{bb}}.$$

The bonds being identical at all three interfaces, the corresponding works of adhesion would follow the same relationship,

$$W_{ab} = \sqrt{W_{aa} \times W_{bb}}.$$

Since the geometric mean of two non-identical values is always less than their arithmetic mean, and because $W_{aa} \neq W_{bb}$ in this case, the relationship among these three parameters would be $W_{bb} < W_{ab} < (W_{aa} + W_{bb})/2$ regardless of the absolute values of n_{ab}, n_{aa} and n_{bb}. Referring to Fig. 3.13, one sees that this set of values is one that will cause the a and b cells to sort out from each other, with the a cell population becoming totally enveloped by the b cell population (Steinberg, 1962c, 1963, 1964) at shape equilibrium.

'Recognition', 'specificity' and 'selectivity' are, after all, imprecise terms. By substituting for such ambiguous phrases as 'selective cell interaction' and 'adhesive recognition' explicit sets of relative W's or σ's characterizing the interfaces in question, and solving for the configurations of lowest interfacial free energy, one can precisely connect a particular set of adhesive relationships with the particular pattern of cell population behaviour which it determines or 'programs.' This has been confirmed with computer models (reviewed in Gordon *et al.*, 1972; discussed in Steinberg, 1975). Thus, for multi-subunit systems that rearrange under the influence of adhesive forces, the equilibrium configurations, analysed thermodynamically, carry information on the relative intensities of adhesion of the various interfaces; and these sets of values in turn may suggest particular categories of physical or chemical mechanisms capable of generating them.

3.2.2 What the DAH explains that 'specific ligands' do not

What shall we say 'causes' the sorting out of the cells depicted in the third row of Fig. 3.15? The ligands at the cell surfaces in this instance are, indeed, specifically interacting, complementary molecular species. However, these are qualitatively identical on the two cell types — but these nevertheless sort out! The ligands show molecular specificity of interaction; because they are present on both cell types, the latter do *not* specifically self-aggregate, but cross-adhere as well; but after adhering to form a mixed aggregate, they sort out within its interior because of differences in the intensities of intercellular adhesion that favour the exchange of neighbours. This exchange continues until that configuration is approached in which the sum of adhesive energies at all cell boundaries within the aggregate is maximized. In this example, *combining-specificity exists at the molecular level, is lost at the level of cell pairs and reappears* de novo *at the cell population level.* Moreover, if we removed the complementary macromolecular ligands and replaced them, one for one, with any other kind of adhesion sites or sources of net attraction, the results would be exactly the same. In fact, all of the tissue configurations depicted in the right-hand column of Fig. 3.15 are shown to be capable of being determined by appropriate distributions of 'specific ligands'. The three configurations at the bottom right belong to categories C and D in Fig. 3.13. The upper one of these — separate masses of dark and light cells — would result from negligible interaction between the 'round' and 'square' binding systems. The two lower figures represent the consequences of allowing 'round' and 'square' binding systems to interact, but rather weakly. In the bottom figure, the dark cells cohere with the same intensity as light cells, whereas in the middle figure the dark cells cohere more strongly. However, a set of W's determining the configuration represented at the right in row 3 of Fig. 3.15 (configuration B in Fig. 3.13) can easily be generated by almost any type of bonding system. It does not require (although it does permit) the use of complementary molecular adhesion sites. In fact, starting with a uniform cell population, even increasing or decreasing the net electrical charge on a subpopulation of the cells, for example by adding or subtracting sialic acid residues to cell surface glycoproteins (Deman *et al.,* 1974), could probably cause the two cell populations to sort out in the configuration shown (see Brick *et al.,* 1974). The ease with which this set of *W*'s can be generated is particularly significant because *this is the type of configuration most commonly observed when vertebrate embryonic cells and tissues sort out from or spread upon one another* (Steinberg, 1970).

The assembly behaviour of any multicellular complex capable of cell rearrangements ultimately depends upon the associative interactions of molecules on the cell surfaces, but not necessarily in a simple, straightforward way. 'Specificity' can either arise anew or be extinguished in the causal chain that translates molecular interactions into cellular adhesions and cellular adhesions into tissue organisation. Molecular interactions (including electrostatic forces of repulsion) cause cells to adhere to one another with particular intensities. These adhesive intensities contribute significantly to cell population σ's. The *set* of σ's at the cell population interfaces in a multicellular system then determines its assembly behaviour. The character of this behaviour and how it results from the adhesive interactions is not always obvious and only emerges from precise and quantitative analysis. Thus, the specific ligand hypothesis does not explain why one cell population sorts out internally to another rather than externally. Invoking tissue-specific ligands and a preference for 'self'-associations, it does not explain why a ball of one tissue will disrupt its own internal 'self'-adhesions in order to spread over and envelop a ball of a 'foreign' tissue. It does not explain why the configuration adopted when one tissue spreads over another is the same as the configuration adopted when their dissociated and intermixed cells sort out from one another. It does not explain why in some combinations this 'equilibrium configuration' is one of *total* envelopment, in other combinations it is one of *partial* envelopment (Steinberg, 1970) and in still other cases it is one of little or *no* envelopment ('self-isolation' in Holtfreter's terminology; see Holtfreter, 1939 and Fig. 3.13). Since the tissue-specific ligand hypothesis explains the 'recognition behaviour' of each tissue by interactions that are unique or specific to it, it cannot explain why one can deduce the internal/external sorting behaviour of cells in untested combinations by reference to their positioning behaviour in other, already-tested combinations (transitive rule; see Section 3.2.1). Neither can it explain why this positioning behaviour correlates with the equilibrium shapes tissue fragments adopt during prolonged centrifugation. These are the things that the differential adhesion hypothesis explains.
These are the things that the differential adhesion hypothesis explains.

As I wrote in 1970, 'The chemistry of cell adhesion and the physics of morphogenetic assembly processes both enter the explanations we are seeking, but in different roles. ... What the chemistry will help us to explain, as it comes to be understood, is why a particular set of adhesive relationships, and not some other set, exists in any particular case. The differential adhesion hypothesis will provide the link connecting the chemistry of intercellular adhesiveness to the morphogenetic behavior of the cell population'.

NOTES

1. Contrary to recent assertions based upon intuitive arguments (Harris, 1976), there are no hidden assumptions, e.g. that the liquid body is energetically a closed system, or that only certain 'reversible' types of bonds are acting to produce the σ, or that the bonds that are broken when an aggregate expands its area must be the same ones or equivalent to those that were established when the aggregate was first formed. 'Reversible' work has nothing to do with any supposed requirement for 'reversible'. bonds; rather, the reversible work done in any process (e.g. lifting a weight) is merely the minimum work by which the process could theoretically be achieved, (with frictionless pulleys, etc.). One cannot *do* reversible work, but one can *measure* it. Thus the *reversible* work required to lift a one-gram weight one centimeter in a gravitational field of 980 cm s^{-2} is 980 ergs. Similarly, the reversible work required to expand by one cm^2 the surface of a liquid body (a body whose σ is area-invariant) with a σ of 20 erg cm^{-2} is 20 erg, *regardless of the bonding mechanism or mechanisms responsible for producing the* σ. The proper measurement of liquid σ's depends upon accurate measurement of the relevant variables and the demonstration that the body in question is truly at shape equilibrium as defined for a liquid − and nothing more.

2. It has been pointed out (Harris, 1976) that because, in the case of cell aggregates, internal adhesions are often focal, occupying only a very small fraction of the area of cell apposition, 'it would be possible for a cell within an aggregate to move from the ... interior to the ... surface without sacrificing any of these ... junctions, assuming that these adhesions could simply 'slide' across the cell surface to the area still juxtaposed to other cells'. This is indeed possible but, contrary to Harris' assertion, it in no way violates any tenet of the DAH, which makes no claims concerning how much reversible work will be required to expand the surface area of any actual aggregate by any particular amount. Aggregates of some kinds do not spontaneously adopt spherical shapes, possibly in some cases because the non-adhesive portions of the surface exceed the area of a sphere of the aggregate's volume. The W required to increase an aggregate's surface area by exposing additional non-adhesive surface would simply be zero unless some other category of reversible work (e.g. to produce a shape-change) were involved.

3. '... findings ... clearly point to qualitative differences between specific constituents on the cell surfaces as being primarily responsible for selective attachment and histogenetic affinities of cells. The facts do not support the notion that quantitative differences exclusively are responsible for the process involved in sorting of cells and multicellular organization (Steinberg, 1963a)' (Moscona, 1968).

4. 'Only quantitative differences in adhesiveness are necessary. ... This does not mean, of course, that molecules of different sorts, on the surfaces either of cells of a given kind or of cells of different kinds, may not in such cases participate directly in the mediation of adhesions. It merely means that whatever the chemical nature of, or diversity among, the adhesives themselves, the quantitative adhesive relationships among the cells which bear them would be expected to approximate, within the limits shown in Fig. 3.2 (similar to Fig. 3.13 in the present paper), ... the relationships derived from the simple postulates which have been outlined. In cases in which, at

equilibrium, one tissue covers the other incompletely or not at all, it becomes *necessary* (emphasis added) to assume ... factors such as an ordered distribution of, or qualitative nonidentity among, adhesive sites' (Steinberg, 1963, p. 407).

And again: 'The latter type of behavior' (incomplete coverage of one cell population by another at configurational equilibrium) 'cannot be obtained from the operation of quantitative differences alone, as in the stochastic model. Other differentials, either of quality or of pattern, must be employed as coding factors for the sites of adhesion' (Steinberg, 1964, p. 350).

REFERENCES

Armstrong, P.B. (1970), *J. Cell Biol.,* **47**, 197–210.

Armstrong, P.B. (1971), *Wilh. Roux' Arch.,* **168**, 125–141.

Barbera, A.J. (1975), *Dev. Biol.,* **46**, 167–191.

Barbera, A.J., Marchase, R.B. and Roth, S. (1973), *Proc. natn. Acad. Sci. U.S.A.,* **70**, 2482–2486.

Brick, I., Schaeffer, B.E., Schaeffer, H.E. and Gennaro, J.F., Jr. (1974), *Ann. N.Y. Acad. Sci.,* **238**, 390–407.

Cassiman, J.J. and Bernfield, M.R. (1976), *Dev. Biol.,* **52**, 231–245.

Crandall, M.A. and Brock, T.D. (1968), *Bact. Rev.,* **32**, 139–163.

Curtis, A.S.G. (1969), *J. Embryol. exp. Morph.,* **22**, 305–325.

Curtis, A.S.G. (1973), *Prog. Biophys. Mol. Biol.,* **27**, 315–386.

Deman, J.J., Bruyneel, E.A. and Mareel, M.M. (1974), *J. Cell Biol.,* **60**, 641–652.

Dupré, '*Théorie Méchanique de la Chaleur*', p. 369, (1869), Cited in Davies, J.T. and Rideal, E.K. (1963), *Interfacial Phenomena*, 2nd edn, Academic Press, New York and London, p. 19.

Gershman, H. (1970), *J. exp. Zool.,* **174**, 391–406.

Gordon, R., Goel, N.S., Steinberg, M.S. and L.L. Wiseman (1972), *J. Theor. Biol.,* **37**, 43–73.

Gottlieb, D.I., Rock, K. and Glaser, L. (1976), *Proc. natn. Acad. Sci. U.S.A.,* **73**, 410–414.

Harris, A.R. (1976), *J. Theor. Biol.,* **61**, 267–285.

Heaysman, J.E.M. and Pegrum, S.M. (1973), *Exp. Cell Res.,* **78**, 71–78.

Henkart, P., Humphreys, S. and Humphreys, T. (1973), *Biochemistry,* **12**, 3045–3050.

Holtfreter, J. (1939), *Arch. exp. Zellforsch. Gewebezucht.,* **23**, 169–209.

Holtfreter, J. (1944a), *Rev. Can. Biol.,* **3**, 220–249.

Holtfreter, J. (1944b), *J. exp. Zool.,* **95**, 171–212.

Humphreys, T. (1967), *The Specificity of Cell Surfaces*, (Davis, B.D. and Warren, L., eds), Prentice-Hall, Inc. Englewood Cliffs, N.J., pp. 195–210.

Lilien, J. and Moscona, A.A. (1967), *Science,* **157**, 70–72.

Marchase, R.B., Barbera, A.J. and Roth, S. (1975), Cell Patterning. *Ciba Found. Symp.* 29 (new series), 315–327, Associated Scientific Publishers, Amsterdam.

McClay, D.R. and Baker, S.R. (1975), *Dev. Biol.,* **43**, 109–122.

McGuire, E.J., and Burdick, C.L. (1976), *J. Cell Biol.*, **68**, 80–89.

Moscona, A. (1961), *Growth in Living Systems*, (Zarrow, M.X., ed.), pp. 197–220.

Moscona, A. (1962), Symposium on Specificity of Cell Differentiation and Interaction, *J. Cell Comp. Physiol.*, **60**, (Suppl. 1), pp. 65–80.

Moscona, A. (1963), *Proc. natn. Acad. Sci. U.S.A.*, **40**, 742–747.

Moscona, A. (1968), *Dev. Biol.*, **18**, 250–277.

Moscona, A. and Moscona, M.H. (1962), *Anat. Rec.*, **142**, 319.

Moyer, W.A. and Steinberg, M.S. (1976), *Dev. Biol.*, **52**, 246–262.

Overton, J. (1962), *Dev. Biol.*, **4**, 532–548.

Overton, J. (1977), *Dev. Biol.*, **55**, 103–116.

Phillips, H.M. (1969), Ph. D. Thesis, The Johns Hopkins University, Baltimore. Md.

Phillips, H.M. and Steinberg, M.S. (1969), *Proc. natn. Acad. Sci. U.S.A.*, **64**, 121–127.

Phillips, H.M. and Steinberg, M.S. (1978), *J. Cell Sci.*, (in press).

Phillips, H.M., Steinberg, M.S. and Lipton, B.H. (1977a), *Dev. Biol.*, **59**, 124–134.

Phillips, H.M., Wiseman, L.L. and Steinberg, M.S. (1977b), *Dev. Biol.*, **57**, 150–159.

Roth, S. (1968), *Dev. Biol.*, **18**, 602–631.

Roth, S., McGuire, E.J. and Roseman, S. (1971), *J. Cell Biol.*, **51**, 525–535.

Roth, S.A. and Weston, J.A. (1967), *Proc. natn. Acad. Sci. U.S.A.*, **58**, 974–980.

Steinberg, M.S. (1962a), *Proc. natn. Acad. Sci. U.S.A.*, **48**, 1577–1582.

Steinberg, M.S. (1962b), *Science*, **137**, 762–763.

Steinberg, M.S. (1962c), *Proc. natn. Acad. Sci. U.S.A.*, **48**, 1769–1776.

Steinberg, M.S. (1963), *Science*, **141**, 401–408.

Steinberg, M.S. (1964), *Cellular Membranes in Development*, (Locke, M., ed.), Academic Press, New York, pp. 321–366.

Steinberg, M.S. (1970), *J. exp. Zool.*, **173**, 395–434.

Steinberg, M.S. (1975), *J. Theor. Biol.*, **55**, 431–444.

Symon, K.R. (1971), *Mechanics*, Addison-Wesley, Reading, Mass., U.S.A.

Townes, P.L. and Holtfreter, J. (1955), *J. exp. Zool.*, **128**, 53–120.

Umbreit, J. and Roseman, S. (1975), *J. biol. Chem.*, **250**, 9360–9368.

Vosbeck, K. and Roth, S. (1976), *J. Cell Sci.*, **22**, 657–670.

Walther, B.T., Ohman, R. and Roseman, S. (1973), *Proc. natn. Acad. Sci. U.S.A.*, **70**, 1569–1573.

Wessells, N.K. (1977), *Tissue Interactions in Development*, W.A. Benjamin, Inc. Menlo Park, California.

Wiseman, L.L., Steinberg, M.S. and Phillips, H.M. (1972), *Dev. Biol.*, **28**, 498–517.

4 Molecular Interactions in Specific Cell Adhesion

J. LILIEN, J. HERMOLIN and P. LIPKE

Specificity of Embryological Interactions
(*Receptors and Recognition,* Series B, Volume 4)
Edited by D.R. Garrod
Published in 1978 by Chapman and Hall, 11 New Fetter Lane, London EC4P 4EE
© Chapman and Hall

4.1 INTRODUCTION

Continuous rearrangements of tissues and cells occur throughout embryogenesis. Initially, these changes are dramatic, creating from a mass of cells the three germ layers, a primitive gut and the embryonic axis. Later rearrangements are more subtle, creating organized cellular patterns within individual tissues. Studies of morphogenesis thus focus on questions of how cells get to their destination. In addition one must consider how cellular patterns are stabilized once achieved.

The foundations for contemporary experimental approaches to morphogenetic rearrangements are based largely on the work of Johannas Holtfreter (1943). He demonstrated that amphibian embryos could be disaggregated into single cells, and that such cells would readhere and sort-out, forming tissue-specific patterns of organization strikingly similar to those achieved *in vivo* (Townes and Holtfreter, 1955). The technique of *in vitro* reaggregation was utilized and refined by many investigators, most notably by Aaron Moscona and Malcolm Steinberg. They characterized the behavior of single cells disaggregated from a variety of embryonic sources and developmental stages when recombined *in vitro* (reviewed by Steinberg, 1964; Moscona, 1965).

For many years a working hypothesis has been that selective intercellular affinities were the basis for *in vitro* sorting-out and that programmed changes in such affinities might be the guiding forces which determined morphogenetic movements *in vivo*. However it was not until 1967 that Roth and Weston (1967) developed a method appropriate for measuring selective affinities. Their assay utilized radio-actively labeled single cells of one tissue mixed with preformed aggregates or tissue pieces from the same tissue and from a second tissue. After a suitable incubation period the number of single cells adherent to each tissue was determined. These initial experiments and a variety of similar analyses demonstrated that single cells preferentially adhere to the homotypic tissue. Such selective intercellular adhesion is now a well established characteristic of embryonic cells (reviewed by Marchase *et al.*, 1976 and Lilien *et al.*, 1977).

On the other hand, it is still not clear that selective cellular affinities are controlling elements in morphogenetic rearrangements. There are, however, several systems in which changes in cell affinities have been temporally correlated with morphogenetic events. Notable are changes in affinity between amphibian ectoderm and endoderm which correlate with invagination of mesoderm (Holtfreter, 1939), and a change in adhesive capacity of teleost blastoderm cells at the outset of epiboly (Trinkaus, 1963). In addition, studies in the chick embryo on the interaction of neural retina cells with the optic tectum provide compelling evidence that morphogenetic potential is reflected in adhesive preferences. Such analysis demonstrates that the pattern of adhesive preferences of cells from dorsal and ventral halves of the retina for dorsal and ventral halves of the tectum mimics the patterns of innervation between these two tissues (Barbera *et al.,* 1973; Barbera, 1975).

An understanding of the role of adhesive interactions in the formation and stabilization of tissue architecture is predicated on molecular characterization of the components of adhesive bonds and a knowledge of how these components interact. Thus a great deal of work has focused on studying the re-formation of intercellular adhesions following dispersal of tissues into single cells. In this review we will concentrate on the molecular basis for selective cell adhesion including a synthesis of the work being carried out in our laboratory.

4.2 EXPERIMENTAL METHODS AND VARIABILITY

Investigations of adhesive interactions among cells have resulted in information that is difficult to integrate and often in conflict. Interpretation of this body of inform-ation depends on analysis of sources of experimental variability. Such variability has ramifications for the characterization of adhesive components. In studies on embryonic cells, those experimental variables which might be of greatest conse-quence are;

(1) embryonic age,
(2) methods used to assess adhesion, and
(3) methods used to prepare single cell populations.

Several reports have shown changes related to cell adhesion with increasing age in chick embryos. Rutishauser *et al.* (1975, 1976) have shown age-dependent changes in adhesive preference among neural retina and brain cells, in addition Gottlieb *et al.* (1974) have reported temporal changes in the effect of neural retina membrane fractions on adhesion of neural retina cells. While the developmental significance of these studies is at present uncertain, they are particularly important as the embryonic ages used span those most often utilized in studies on cell adhesion.

Early studies by Moscona (1962) have also shown that, with increasing developmental age, dissociated embryonic cells show a decreased ability to form compact aggregates. This decrease is accompanied by an apparent decrease in amount of membrane-associated aggregation enhancing factor and a decrease in the ability of single cells to respond to the factor (Hausman and Moscona, 1976). This is consistent with a diminished role for the factor with increased age. The progressive increase in frequency of desmosomes during development (Hay and Revel, 1969), presumed to play a role in stabilization of tissue architecture, also implies that qualitatively differing types of adhesive interactions come into play as development proceeds.

Differences in assays used by various laboratories are major and the results obtained are correspondingly different. For instance, the readhesion of populations of single cells has been assessed both by examining the rate of adhesion immediately following dissociation and by examining the size of compact reaggregates formed over a 24 hour period. These two phases of readhesion appear to be reflections of distinct types of interactions. This was first suggested by the apparently random initial adhesions seen when cells from differing tissues were mixed (Trinkaus and Lentz, 1964), whereas homotypic adhesions predominate following several hours of culture (Steinberg, 1964). This difference is underscored by comparing the effects of perturbations in both phases. The initial adhesive interaction of trypsin-dispersed single cells has been shown to be largely non-specific (Roth, 1968), independent of protein synthesis (see Lilien, 1969) and independent of divalent cations (Grunwald and Lilien, unpublished); while those adhesions involved in the formation of compact reaggregates manifest specificity and are dependent on both protein synthesis (Moscona and Moscona, 1962) and divalent cations (Moscona, 1962).

More recently, Umbreit and Roseman (1975) have shown that initial adhesion may be further subdivided into at least two distinct types following single cell preparation with pronase: unstable adhesions which are insensitive to KCN and stable adhesions which are sensitive KCN. Of particular interest is the fact that similar results were not seen when the cells were dispersed using trypsin. This difference underscores the importance of the dissociation method on the apparent mechanism of adhesion.

While there has been no systematic study of differing dissociation methods on the rate of single cell adhesion among embryonic cells, it is clear that trypsin concentration does have an effect (McQuiddy and Lilien, 1971; Morris, 1976). Adhesion among established cell lines of fibroblasts also reflects similar phenomena. Edwards *et al.* (1975) have reported that sensitivity of adhesion to inhibitors of protein synthesis is dependent on the amount of trypsin used in preparation of single cells. In addition, dependence of adhesion of BHK cells on divalent cations appears to vary depending on the degree of trypsinization (Edwards *et al.,* 1975; Urushihara *et al.,* 1976).

Another widely used technique for measuring adhesive interactions measures the rate of single cell adhesion to preformed cellular aggregates or monolayers (see Marchase *et al.,* 1976). Such assays show varying degrees of specificity and susceptibility to inhibitors of protein synthesis when used with embryonic cells. Such differences occur even when the cell type is constant: McClay and Baker (1975) have shown little specificity and complete inhibition by cycloheximide while Roth *et al.* (1971a) have shown a high degree of specificity and little effect of inhibitors of protein synthesis. The source of these differences is not clear, but may be due to differences in the ratio of collecting surface to single cell surface and/or in the preparation of single cells. McGuire and Burdick (1976) have examined the effect of dissociation procedure on collection of cells by aggregates and have found tremendous variation in rate depending on both the concentration of trypsin and the time of proteolysis.

This brief account points to possible difficulties in attempting to integrate data obtained utilizing differing cells and techniques. Thus many of the apparent differences in adhesive interactions may be due to assays which measure different phases in the re-formation of adhesive bonds. These phases may be artifacts generated by cell populations which are stripped of their adhesive components to varying degrees during dispersal and have replaced them to different extents. Collecting surface assays utilize a freshly dispersed population of single cells and a surface of cells which have undergone extensive repair which sets these assays apart.

4.3 ADHESIVE COMPONENTS AND THEIR ASSAY

Several laboratories have attempted to isolate and characterize components of the adhesive mechanisms. Variations in the experimental systems and assays have resulted in apparently differing factors, possibly reflecting different types of interactions during the formation of an adhesive bond.

4.3.1 Assays based on aggregate size

Efforts to isolate components of the adhesive mechanism were initiated by Moscona (1962). He reasoned that cells prevented from readhering might synthesize adhesive components and release them into the external medium via normal turnover processes. Media conditioned by embryonic chick neural retina cells were assayed for their effect on aggregation by comparison of the size of aggregates formed in its presence to controls after 24 h of rotation culture. These experiments demonstrated that the conditioned media contained a non-dialyzable component which increased aggregate size.

Lilien and Moscona (1967) and Lilien (1968) prepared media conditioned by monolayer cultures of embryonic chick neural retina cells. The monolayers were prepared in the presence of serum, subsequently washed and then incubated with serum-free medium. The conditioned, serum-free medium contained a tissue-type-specific component(s) which also enhanced aggregate size. Garber and Moscona (1972) have extended these studies by demonstrating that a tissue-specific factor can be prepared from embryonic mouse cerebrum cells in monolayer.

Work has continued in Moscona's laboratory on the purification and characterization of the neural retina aggregation enhancing component. Hausman and Moscona (1975) have reported that the purified component is a glycoprotein of approximately 50 000 daltons. Of relevance when comparing this factor with others to be described is that the carbohydrate is not required for activity.

More recently Hausman and Moscona (1976) have reported that the aqueous phase from butanol extracts of membrane preparations isolated from neural retina contain a component with similar activity and characteristics (see also Hausman *et al.*, 1976). Activity could not be extracted from embryos older than 13 days nor did cells older than 13 days respond to the factor. These data suggest some form of temporal regulation of both accumulation and response.

Aggregation enhancement appears to result from the progressive enhancement of the size of clusters, beginning at about one hour after the onset of aggregation (Hausman and Moscona, 1976). These data imply that the factor does not affect the initial rate of aggregation but requires a period of metabolic activity. Although the authors have considered this factor to be an intercellular ligand, the data are equally compatible with its having a catalytic role or a role in stimulating endogenous synthesis of adhesive components.

4.3.2 Kinetic assays

Kinetic assays for cell adhesion have been used to examine the process of aggregation during the first 2 hours or less following single cell preparation. These assays are usually quantitated by monitoring the disappearance of single cells on a particle analyzer (Ball, 1966; Lilien, 1968; Orr and Roseman, 1969; Edwards and Campbell, 1971) or by changes in absorbance which accompany progressive aggregate formation (Cunningham and Hirst, 1967).

Merrell and Glaser (1973) have utilized such an assay to characterize the effect of plasma membrane preparations on cell adhesion. Using cells and membranes from chick embryo neural retina and cerebellum they showed that adhesion was specifically inhibited by homologous membranes. In addition membranes were shown to bind only to homologous cells. Interestingly, binding was shown to be temperature dependent. This observation suggests that binding requires the synthesis of a cell surface component or some other temperature dependent surface property, such as lateral mobility of receptors. Gottlieb *et al.* (1974) have also shown that, in addition to tissue specificity, there is temporal specificity. Membranes isolated from 7, 8 or 9 day retina are most effective in inhibiting adhesion among cells of the same age. The authors also found that retinal cell membranes inhibited adhesion of tectal cells and that the same age dependency prevailed, that is, membranes from 8 or 9-day retina were most effective on tectal cells of the same age. The striking temporal specificity reported indicates that rapid developmental changes in surface specificity can occur.

Initial attempts at isolation of the active membrane component have resulted in 20 fold purification of the retina factor. The purification involved delipidation of the membranes with acetone, extraction of the activity with lithium diiodosalicylate and filtration through molecular sieve membranes (Merrell *et al.*, 1975). The active fraction had an apparent molecular weight of 60 000 as determined by polyacrylamide gel electrophoresis in sodium dodecyl sulfate.

The importance of carbohydrate in cell adhesion has been investigated in BHK cells by Vicker (1976). He trypsinized cells and pronase-digested the soluble fraction. The resulting glycopeptides were subsequently separated by high voltage electrophoresis and tested for their effect on the rate of cell adhesion. Two fractions were found to inhibit cell adhesion. The activity of these glycopeptides was destroyed by treatment with periodate or galactose oxidase while the activity of one fraction was enhanced by neuraminidase. This enhancement suggests that some of the active residues are masked *in situ* by terminal sialic acid residues. This is consistent with enhanced adhesion of BHK 21 cells following neuraminidase treatment (Vicker and Edwards, 1972). These studies clearly implicate carbohydrate, specifically galactosyl residues, in BHK cell adhesion.

Inhibition of cell adhesion by membranes or glycopeptides is most easily visualized in terms of direct surface interaction of two complementary molecules; i.e. the active fraction competes with cells for their complementary grouping. While this is the simplest interpretation other alternatives cannot be excluded.

On the other hand, Oppenheimer and Humphreys (1971) have shown that mouse ascitic fluid in which teratoma cells were grown stimulates adhesion of teratoma cells. This fluid is inactive on several other cell types tested, indicating some specificity. The factor preparation is active on glutaraldehyde-fixed cells at $37°C$ but is inactive at $4°C$. More recently Oppenheimer (1975) has demonstrated that treatment of the factor with β-galactosidase abolishes activity. In addition, galactose, galacturonic acid, glucuronic acid and neuraminic acid are effective inhibitors of adhesion. Meyer and Oppenheimer (1976) have reported that at least two separable components in the ascitic fluid are essential for activity.

Studies on sponge cells have shown that isolated macromolecular components promote specific adhesions in the absence of metabolic activity (Humphreys, 1963). While the original aggregation factor prepared by Humphreys (1963) has been purified and appears to be a single component (Cauldwell *et al.*, 1973; Henkart *et al.*, 1973),Weinbaum and Burger (1973) have presented data suggesting that a second component termed 'baseplate' may also be required. Sugar residues have also been implicated in the adhesion of sponge cells (Turner and Burger, 1973). A significant difference between the sponge and teratoma systems is the apparent temperature dependence of the teratoma factor which suggests that a catalytic function or conformational change is necessary for activity.

4.3.3 Binding of cells to treated nylon fibers

Rutishauser *et al.* (1975, 1976) developed an assay which measures the binding of live cells to nylon fibers coated either with immobilized cells or with antibodies directed against proteins of conditioned culture media. Using this assay procedure they have shown that chick embryo neural retina cells release into culture medium, 2 proteins of molecular weight 55 000 and 140 000 which may be involved in adhesion. The larger molecule may be cleaved by proteolysis to yield a fragment similar or identical to the smaller molecule. Both retina and brain cells were shown to bind to nylon fibers coated with antibodies to the smaller molecule. However binding differed for the two cell types as a function of embryonic age. The binding of either cell type to cells attached to fibers was inhibited by antibodies directed against the smaller molecule, suggesting a role for this molecule in cell adhesion.

4.3.4 Binding of macromolecules to cells

Several years ago we became convinced of the necessity to develop new approaches to identify the number and types of components involved in specific cell adhesion. Our reasons were based on several considerations. Assays measuring aggregate size do not allow one to distinguish direct participation in the formation of adhesions from a variety of indirect effects, nor do they allow for descriptions of temporally distinct events. In addition the kinetics of cell adhesion, at least for trypsin-dispersed embryonic cells, appear to be complicated by non-specific interactions which may obscure specific events. Our first approach was based on the early findings of Lilien (1968) that the neural retina aggregation enhancing component was specifically adsorbed by homologous cells at 4°C. This prompted us to examine the binding of radioactively labeled aggregation enhancing supernatants to 10-day neural retina cells at 4°C (Balsamo and Lilien, 1974a, 1975). Our initial results were quite encouraging: serum-free organ culture-conditioned media contained macromolecules which bound to cells tissue-type specifically (Balsamo and Lilien, 1974a).

These studies were continued in an effort to identify the moieties essential for binding (Balsamo and Lilien, 1975). When organ culture-conditioned media labeled with both ^{3}H-glucosamine and ^{14}C-leucine were used, both isotopes showed tissue-specific, saturable binding. Following pronase digestion only glucosamine label bound to cells. Digestion with Rhozyme—a crude preparation of glycosidases with little or no detectable protease activity (Bahl and Agrawal, 1972)—abolished binding of both labels. These results suggested that carbohydrates are the essential determinants of binding.

To identify the specific sugars essential for binding three types of studies were performed;

(1) Sugar competition — cells were pre-incubated with monosaccharides prior to binding of radioactive conditioned media,
(2) Sugar stabilization — addition of unlabeled organ culture-conditioned media following binding has been observed to increase the half life of bound radioactive material by a factor of two and therefore we speculated that the appropriate sugar might have the same effect,
(3) Specific glycosidase digestion — labeled organ culture-conditioned media were digested with Rhozyme. The treated media were assayed for binding and the liberated sugars were identified.

The result of these experiments were in complete agreement and indicated that binding of the factor from neural retina-conditioned media was dependent on a terminal *N*-acetylgalactosamine residue and that from cerebral lobe-conditioned media was dependent on a terminal mannosamine-like residue. These binding components recovered from organ culture-conditioned media will be referred to as ligands.

That ligand binding is physiologically significant is shown by studies conducted on both neural retina and optic lobe cells (Balsamo *et al.*, 1976). In these studies radioactively labeled organ culture-conditioned media from optic lobes were shown to bind to both optic lobe and neural retina cells. The binding to neural retina cells was regionally specific, mimicking the normal pattern of innervation. Conditioned media from dorsal, ventral, anterior or posterior optic lobe halves bound preferentially to that quadrant of the retina which normally innervates that region of the tectum. These studies compliment those of Barbera *et al.* (1973) and Barbera (1975) which demonstrated adhesive preferences of dorsal or ventral neural retina cells for the appropriate tectal region and those of Gottlieb *et al.* (1974) which demonstrated inhibition of tectal cell adhesion by retinal membranes. The concurrence between these three widely differing approaches to studying cell interactions (see also Lilien *et al.*, 1977 for further discussion) suggests that the methods assess similar phenomena.

4.3.5 Agglutination assays

The binding studies established that tissue-type-specific ligands are released from intact tissues in culture; however they did not establish a role for ligand in intercellular adhesion. We attempted to show such a relationship by demonstrating an effect of ligand-exposed cells on the size of aggregates formed by freshly dispersed cells (Balsamo and Lilien, 1974b). Because of the rapid reversibility of ligand binding, the cells were fixed with glutaraldehyde following binding. Aggregation assays were performed at various ratios of fixed to live cells. Fixed cells which had not been exposed to ligand had no effect on aggregation while those exposed to ligand caused a dramatic increase in aggregate size. The effect was specific, only homologous fixed cells were capable of stimulating aggregation.

Enhanced aggregation by fixed cells was dependent on protein synthesis, suggesting that the interaction of live cells with the fixed cells required the synthesis of an additional component. We reasoned that this component might be synthesized and released into the environment where it could be effective in agglutinating the fixed cells. This hypothesis was tested by incubating fixed cells above a monolayer prepared in serum-free medium. The cultures were rotated at 70 rev/min to maintain the fixed cells in suspension and incubated at 37°C for 24 hours. The fixed cells agglutinated if, and only if, the monolayer was of a homologous type and the cells had been exposed to homologous ligand prior to fixation. The component contributed by the monolayer is not ligand since ligand-containing solutions do not agglutinate fixed cells. Therefore monolayers must produce a distinct component which we have called a ligator or an agglutinin.

These results are significant in that they demonstrate that a specific adhesive bond can be reconstituted from distinct soluble components in the absence of cellular metabolism. It is therefore possible to isolate and study the individual components and to reconstitute the system from its component parts.

The agglutinin is stabilized by addition of protease inhibitors to the freshly collected monolayer-conditioned media. Its activity is quantitated by measuring the remaining single cells following a suitable incubation period (Lilien and Rutz, 1977). Using this assay we have evaluated several parameters of the agglutination system. Agglutination is both time- and temperature-dependent. Maximal agglutination is achieved only after 18 h at 37°C (Lilien and Rutz, 1977). In contrast, lectin-mediated agglutination of retina cells is rapid and maximal at room temperature (McDonough and Lilien, 1975).

The temperature dependence exhibited by this system is not comparable to that of most other adhesive systems. The temperature dependence of aggregate size enhancement (Hausman and Moscona, 1976) and of membrane binding Merrell and Glaser, 1973) is most easily interpreted as an effect on cellular activities. In the case of agglutination of fixed retina cells and in the case of the teratoma factor activity of the component itself is temperature-dependent.

Activity of the agglutinin is dependent on divalent cations (Lilien and Rutz, 1977); removal of Ca^{2+} and Mg^{2+} reversibly inactivates it. Interestingly, while the sponge factor also requires Ca^{2+}, removal results in irreversible inactivation (Humphreys, 1967). At this point it is impossible to distinguish whether the agglutinin requires divalent cations for its stability or integrity, or whether they are required for interaction with other molecular species. In either case ion dependency implies that agglutination is a reflection of live cell adhesion.

Agglutination requires the interaction of cells with organ culture-conditioned medium (ligand) followed by agglutination in the presence of monolayer-conditioned medium (agglutinin). That these two sources appear to contain different components is not surprising. The agglutinin is extremely labile to serine proteases. Upon collection from monolayers in the absence of inhibitors activity is lost almost immediately (Rutz and Lilien, unpublished). That organ culture-conditioned media do not have a similar activity is reasonable, since within a solid tissue diffusion is probably limited, resulting in inactivation by cell surface proteases. The absence of ligand in monolayer-conditioned medium may be explained by the difference in cell concentration. Organ cultures contain $\simeq 5 \times 10^7$ cells ml^{-1}, while monolayers contain only one tenth of that cell density. Thus the absence of ligand activity in monolayer-conditioned medium may be explained simply on the basis of its being present in too low a concentration to be assayed.

Based on our initial agglutination studies (Balsamo and Lilien, 1974b) we had developed a model of the adhesive mechanism consistent with our data on neural retina and cerebral lobe cells (Fig. 4.1a). This model, although not the only possibility (see Fig. 4.1b,c), has allowed us to make simple, testable predictions. It should be noted that any of the individual idiograms of Fig. 4.1 may actually consist of more than one component.

The models of Fig. 4.1 successfully predicted that free ligand would inhibit agglutination of cells by competing for ligand binding sites on the agglutinin. The inhibition is tissue-type-specific: cerebral lobe ligand does not compete in the retina system and *vice versa* (Lilien and Rutz, 1977).

RECONSTRUCTIONS OF THE ADHESIVE BOND

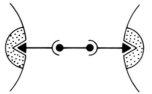

(a) THREE-COMPONENT: RECEPTOR, LIGAND, AGGLUTININ

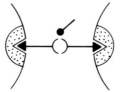

(b) THREE-COMPONENT: RECEPTOR, LIGAND, CATALYST

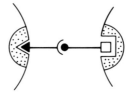

(c) FOUR-COMPONENT: RECEPTOR, LIGAND, LIGAND', RECEPTOR'

Fig. 4.1 Three possible reconstructions of the adhesive bond based on fixed cell agglutination experiments.

The interaction of neural retina ligand and agglutinin was further characterized using enzyme-digested ligand preparations. β-N-Acetylhexosaminidase-digested, but not pronase-digested, ligand preparations were effective competitors of agglutination. In addition, each of these preparations was tested for its ability to mediate agglutination by direct incubation with cells before fixation and assay (Fig. 4.2). In this case, neither preparation was active. These data are most easily interpreted in the following way: glycosidase-digested ligand is unable to bind at the cell surface and thus cannot mediate agglutination. However it has an intact agglutinin-binding site and can compete for the agglutinin. Pronase-digested ligand, although it may bind at the cell surface, can neither mediate nor compete in agglutination, as it lacks the necessary binding site.

We have reviewed data on several adhesive factors from chick neural retina. There is no obvious model which can accomodate all of these factors, indeed they may even participate in different types of adhesions. There are however, provocative relationships between several of the preparations. The factor originally described by Lilien (1968), that purified by Hausman and Moscona (1975) and the agglutinin described by Lilien and Rutz (1977) are all obtained from monolayer-conditioned

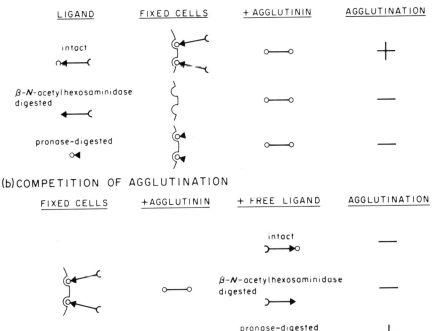

Fig. 4.2 Representation of fixed cell agglutination experiments based on Model a of Fig. 4.1.

medium and require an extended time-course for activity. The membrane fraction prepared by Merrell *et al.* (1975) may be a receptor for the tissue-specific ligand. Binding of ligand to the cell surface does not require divalent cations (Lilien and Rutz, 1977), nor are divalents required for interaction of the membranes with cells (Merrell *et al.,* 1976). While ligand from optic lobes interacts with neural retina cells the membrane factor from neural retina interacts with optic lobe cells (Gottlieb *et al.,* 1974). Neither of these interactions occur in the reverse direction. Final integration of these factors will depend on cross-testing the various components in the several reported assays.

4.4 CONTROL OF CAPPING OF SURFACE RECEPTORS

4.4.1 A macromolecular inhibitor of capping

Direct binding assays and agglutination assays reveal that neural retina and cerebral lobe organ culture-conditioned media contain ligands which bind to cells through saccharide moieties. We have also reported that organ culture-conditioned media inhibit the induced capping of cell surface receptors for three different plant lectins (McDonough and Lilien, 1975a). This activity is also tissue specific and, like binding, is labile to β-N-acetylhexosaminidase or α-mannosidase for retina or cerebral lobe cells respectively. The dependence of each of these assays on the same terminal sugar suggests that a similar or identical ligand is responsible. Inhibition is not due to competition for lectin binding sites. This is indicated by the tissue specificity of the ligands and their ability to inhibit capping of receptors for several lectins with different carbohydrate specificities. Furthermore, the amount of fluorescent lectin bound is unaltered by the presence of the ligand. The mechanism of this inhibition does not appear to be similar to high Concanavalin-A (Con A) inhibition of immunoglobulin receptor capping observed in lymphocytes (Yahara and Edelman, 1972; Edelman *et al.*, 1973). Unlike the latter system, colchicine does not reverse the effect. Our data are consistent with two general ideas:

(1) inhibition mediated by a generalized structural change within the membrane itself and/or
(2) an induced alteration of cytoskeletal elements, possibly a loss of their membrane attachments.

Two observations imply that capping inhibition by ligand is physiologically significant. First, only fresh, trypsin-dispersed cells are able to cap; cells within intact retinas or cells which have been allowed to repair in culture for 2—4 hours are incapable of capping (McDonough and Lilien, 1975). This suggests that trypsinization removes a component from the cell surface which normally restricts the flow of membrane receptors into caps. Presumably that component is ligand. Second, organ culture-conditioned media from tectal halves inhibit capping of retina cells from each of the four quadrants. The results exactly mimic binding; inhibition parallels the normal pattern of innervation (Balsamo *et al.*, 1976).

4.4.2 Turnover of the capping inhibitor

We have examined the parameters governing the loss of capping ability of neural retina cells during repair in culture (McDonough *et al.*, 1977, McDonough and Lilien, 1977). When 10-day retina cells are repaired in the presence of cycloheximide this loss occurs for only 2 h and is followed by a gradual recovery of capping ability. Concomitant with recovery of capping ability, capping inhibitory activity accumulates in the medium. This soluble activity is tissue-specific and labile to

β-*N*-acetylhexosaminidase, suggesting an identity with ligand. One interpretation of these results is that there exists a pool of capping inhibitory ligand which is displaced to the cell surface during repair and subsequently shed into the environment. If this hypothesis is correct cells which have been repaired in cycloheximide, and which have released their ligand into the medium should, upon reculture in cycloheximide, show no loss in capping ability since ligand synthesis is blocked and the pool has been depleted. This is in fact the case. However, if cycloheximide is omitted during the second culture period, there is a gradual loss of capping ability. This suggests that loss of capping ability may result from either mobilization of a pool or *de novo* synthesis when the pool is absent.

Mobilization of ligand from a precursor pool to its site of action is inhibited by colchicine, cytochalasin B, azide and reduced temperature (McDonough *et al.*, 1977). While these data are consistent with a model involving fusion of an intracellular vesicle-enclosed pool with the plasma membrane (Redman *et al.*, 1975; Butcher and Goldman, 1974) they are also consistent with a model involving a cryptic cell surface pool whose availability is dependent on lateral mobility within the membrane. All of the conditions which inhibit pool mobilization also affect the lateral redistribution of cell surface receptors among neural retina cells (McDonough and Lilien, 1975b).

An additional point is that if cycloheximide is omitted during the initial 2 h repair period, reculture in the presence of cycloheximide results in no loss of capping ability. This suggests that the existence of the capping inhibitor as a pool is incompatible with synthesis. Furthermore organ culture-conditioned media (ligand) inhibit pool mobilization (McDonough *et al.*, 1977). This effect is not observed for terminally deglycosylated ligand or other protein solutions. This suggests that there is a feedback mechanism which controls both synthesis and the relative amounts of pool and surface-localized ligand.

The ability of cells to cap lectin receptors following release of endogenous ligand has allowed us to examine the process of release (McDonough and Lilien, 1977). This was facilitated by the observation that, in addition to the slow release of ligand beginning at 2 h during repair in cycloheximide, rapid release could be induced by transfer of the cells to fresh medium at 2 h. Within 10 minutes after transfer the cells completely regain their ability to cap lectin receptors and maximal ligand activity is found in the medium. For these experiments, trypsin-dispersed single cells were allowed to repair in the presence of cycloheximide for 2 h, and were then transferred to fresh medium containing cycloheximide and any effector to be tested for 10 minutes (release period). Following this procedure there is always a direct relationship between the ability of the cells to form caps and the amount of ligand activity found in the medium.

Conditions which inhibit ligand release are reduced temperature or the presence of cytosine arabinoside or hydroxyurea, both in the range of 1–2 mM. Noteworthy is the lack of inhibition by drugs which inhibit pool mobilization (colchicine, cytochalasin B). Since both cytosine arabinoside and hydroxyurea have been reported to inhibit glycosyl transferases (Hautrey *et al.*, 1974) we pursued the notion that release was mediated by terminal glycosylation of the ligand. Consistent with the ion requirements of known glycosyl transferases, EDTA completely blocked release and Mn^{2+}, but not Ca^{2+} or Mg^{2+}, stimulated release (McDonough and Lilien, 1977).

The inhibition of some glycosyl transferases by nucleotides prompted us to test the effect of these compounds on release. Strikingly, of all the nucleotides tested on neural retina cells UDP was the most effective inhibitor, completely blocking release at a concentration of 0.1 μM (McDonough and Lilien, 1977).

These studies on single cells were extended to include the effects of nucleotides on ligand release in organ culture (Hermolin and Lilien, unpublished). Of a number of nucleotides tested, UDP was again the most effective inhibitor of accumulation of activity in retina cultures and GDP was most effective in cerebral lobe cultures. These are the nucleotides which activate N-acetylgalactosamine and mannose respectively for known glycosyl transferases. Thus the nucleotide specificities match the previously identified terminal sugars essential for inhibition of capping: N-acetylgalactosamine for the neural retina ligand and a mannose-derived residue for the cerebral lobe ligand (McDonough and Lilien, 1975a).

The importance of terminal glycosylation in ligand release was also tested by attempting to reactivate (i.e. reglycosylate) ligand solutions previously inactivated by digestion with β-N-acetylhexosaminidase (McDonough and Lilien, 1977). Three cell preparations were incubated with digested ligand: freshly trypsinized cells, cells which had been repaired and stripped of their endogenous ligand, and cells that had been repaired, stripped and briefly trypsinized. The repaired and stripped cell preparation alone was able to reactivate the enzyme-digested ligand, indicating that the putative transferase is trypsin-labile. Reactivation was completely blocked by UDP. Activity of the reactivated ligand was tissue-type specific and labile to β-N-acetylhexosaminidase indicating its identity with the original preparation. These data are consistent with the hypothesis that release is mediated by terminal glycosylation, and that the pool of ligand most likely exists in a form lacking the terminal sugar.

The interaction of enzymatically deglycosylated ligand with transferase and the acquisition of surface transferase activity during repair in cycloheximide suggested that the pool of ligand may exist in association with the appropriate transferase. To further test this notion we examined the effect of cytochalasin B, ligand solutions (both of which inhibit mobilization of the pool) and enzymatically deglycosylated ligand solutions (which do not inhibit pool mobilization) on the acquisition of surface transferase activity. Of the three, cytochalasin B and native ligand inhibited acquisition of transferase activity (Table 4.1). Thus mobilization of the transferase is governed by rules similar to those that govern mobilization of ligand. This finding supports the notion of an intimate association.

The regulation of pool mobilization by exogenously added ligand suggested that there are additional levels of control. We therefore tested the effect of exogenously

Table 4.1 Effects of culture in cytochalasin B or ligand-containing solutions on sensitivity of cells to native and deglycosylated ligand.

Treatment during culture	% Inhibition of capping by ligands	
	Native	Deglycosylated
None	10	59
Cytochalasin B (40 μg ml^{-1})	59	22
Ligand-containing tissue culture supernatant (75% v/v)	55	24
β-N-acetylhexosaminidase-digested ligand solution (75% v/v)	7	56

Freshly dissociated 10-day chick neural retina cells were cultured for 2 h at 37°C in medium containing cycloheximide (5 μg ml^{-1}) and the additions above (for culture conditions see McDonough and Lilien, 1977). After 2 h the cells were harvested, incubated for 10 min at 37°C in 0.01 M Hepes-buffered 0.15 M NaCl (pH 7.4) containing 2 mg ml^{-1} glucose and 5 μg ml^{-1} cycloheximide (HBSG−CH), and assayed for their sensitivity to glycosylated ligand and deglycosylated ligand solutions (80% v/v). Values shown are means of duplicates from two experiments.

added ligand on the release process. Limiting amounts of either native or β-N-acetylhexosaminidase-digested ligand were added during the release period. Following a 10 minute incubation the cells were pelleted and the supernatant solutions assayed for their effect on capping of fresh, trypsin-dispersed cells. If release were inhibited only activity due to the exogenously added ligand would be present; if, on the other hand, release occurred activity would be increased. The results are illustrated in Fig. 4.3. Of the two preparations tested, only the undigested ligand inhibited release. Thus, there appears to be a series of feedback controls exerted by fully glycosylated ligand both on release of endogenously mobilized ligand and on mobilization of the pool of ligand.

4.4.3 Ligand receptor dynamics

The data presented thus far imply that the ligand pool exists in a terminally unglycosylated form. The loss of capping ability during repair, therefore, implies that this form of the ligand is also active in hibiting capping. As predicted, enzymatically deglycosylated ligand is active in inhibiting capping but only among repaired cells stripped of endogenous ligand. As is the case for transferase activity, a brief trypsinization of these cells abolishes the ability of the deglycosylated ligand to inhibit capping (McDonough and Lilien, 1977). This trypsin lability explains the unresponsiveness of freshly trypsin-dispersed cells to deglycosylated ligand. Thus, as we can link the effect of the fully glycosylated ligand with the existence of a specific cell surface receptor, we may link the effect of the deglycosylated form of

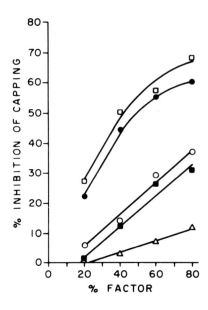

Fig. 4.3 Effects of ligand-containing solutions on release of cell surface material from 10-day chick neural retina cells. Cells were cultured for 2 h at 37°C in the presence of cycloheximide, harvested, and washed with ice-cold Hepes buffered saline with 5 μg ml^{-1} cycloheximide (HBSG-CH; 0.01 M Hepes, 0.15 M NaCl, 1 mg ml^{-1} glucose). After resuspension in HBSG-CH containing various additives the cells were incubated for 10 min at 37°C to release bound ligand, pelleted at 5°C for 10 min at 200g, and the medium collected and assayed on freshly dissociated 10-day neural retina cells for its affect on Con-A induced capping. Values shown are means from duplicates of 2 separate experiments. ●, Release period in HBSG-CH; ○, release period in HBSG-CH + ligand-containing organ culture supernatant (33% v/v); □, release period in HBSG-CH + deglycosylated ligand solution (33% v/v, deglycosylated by treatment with β-N-acetylhexosaminidase); ■, ligand-containing culture supernatant alone (33% v/v); △, deglycosylated ligand solution alone (33% v/v).

the ligand with the existence of a second cell surface receptor, presumably the transferase. That these two receptor populations are distinct is indicated by the inability of the fully glycosylated ligand to affect repaired and stripped cells. We shall refer to the receptor for fully glycosylated ligand as GL receptor and that for the terminally deglycosylated ligand as DL receptor.

Is there a relationship between these two receptor populations? As a first approach to this question we examined the ability of the fully glycosylated ligand and its terminally deglycosylated form to inhibit capping among cells over a six hour period in culture (McDonough and Lilien, unpublished. That ligand in either form

MODEL OF LIGAND TURNOVER

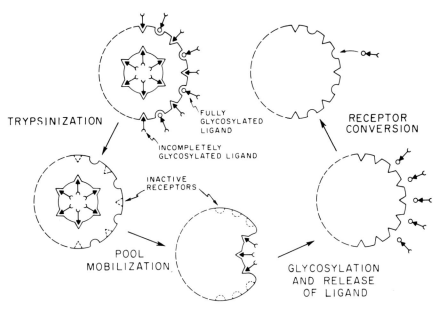

Fig. 4.4 A model of ligand turnover.

inhibits capping implies the presence of the appropriate receptor. Therefore, susceptibility to ligand is an assay for functional, but not physical, presence of receptor.

Following single cell preparation with trypsin there is a gradual increase in DL receptors reaching a plateau at 2 h; concomitantly there is a loss of GL receptors approaching undetectable levels by 2 h, followed by a gradual increase with a maximum at approximately 6 h. Thus it appears that both types of receptors may exist simultaneously. In the presence of cycloheximide the profiles are similar, with the important exception that, beginning at 2 h and continuing until 6 h, there is a gradual loss of DL receptors. The reappearance of GL receptors in the presence of cycloheximide shows that reappearance is not due to new protein synthesis. The inverse relationship exhibited by these two receptor populations beginning at 2h in the presence of cycloheximide suggested the possibility that DL receptors were being converted to GL receptors. If such a conversion were the basis for reappearance of GL receptors one would expect that destruction of DL receptors at 2 h (with trypsin) would result in the failure of the GL receptors to reappear. This is indeed the case; eliminating the possibility that reappearance of GL receptors is due to either cyclic variations in activity or the delayed insertion of a pool. Thus the initial loss of functional GL receptors is a permanent one although it is not possible to determine if it is due to the physical loss of the receptor.

Conversion of DL to GL receptors is the most likely hypothesis to explain these data; however, the reverse conversion appears unlikely under similar conditions. When cells which have both receptors are subjected to a brief trypsinization they do not reacquire sensitivity to the deglycosylated ligand on continued culture in the presence of cycloheximide.

Fig. 4.4 summarizes a scheme for the mobilization, turnover and dynamics of receptor conversion based on exploitation of capping inhibition as a probe. We presume that within an intact tissue both DL and GL receptors are present. The ability of cells to cap attests to the lack of surface ligand following trypsinization of the tissue. There does exist an endogenous pool of ligand and transferase. The data suggest an intimate association between the two, possibly within vesicles. The pool is mobilized to the cell surface, perhaps by the fusion of the vesicles with plasma membrane. After terminal glycosylation of the ligand and subsequent release there is a conversion of the DL receptors, i.e. transferase, to the GL form, which recognizes the terminally glycosylated ligand. Glycosylated ligand interacts with GL receptors manifesting a series of controls over further pool mobilization, and terminal glycosylation (release). While this scheme is consistent with many experimental observations it should be stressed that it is a hypothesis which requires substantiation. It does, however, provide a foundation for directing further analysis. Many of the uncertainties are amenable to experimental verification.

4.5 RELATION OF CAPPING INHIBITION AND AGGLUTINATION

In each of the assays developed in our laboratory: binding, agglutination and capping inhibition, activity of the organ culture-conditioned media is abolished by the same glycosidase; β-*N*-acetylhexosaminidase in the case of retina and α-mannosidase in the case of cerebral lobes. Enzyme lability is suggestive of, but does not prove, identity of the molecular species active in each assay. The identity of these activities has been confirmed for retina cells by two additional types of studies:

(1) acquisition and loss of agglutinability by cells under the same conditions detailed for mobilization and release of the capping inhibitory ligand and
(2) the effect of nucleotides on the accumulation of agglutination-mediating activity in organ culture.

During repair in culture retina cells become agglutinable by monolayer-conditioned medium with the same kinetics as they lose capping ability (Lilien and Rutz, 1977). Acquisition of agglutinability is inhibited by the same conditions which inhibit mobilization of the pool of capping inhibitory ligand, namely reduced temperature, colchicine and cytochalasin B (McDonough *et al.,* 1977). Release of the capping inhibitor from the cell surface is accompanied by a loss of agglutinability (McDonough) *et al.,* 1977). Additionally, the same compounds which inhibit release of the capping inhibitor (cytosine arabinoside, hydroxyurea, EDTA and most significantly UDP) also inhibit the loss of agglutinability (McDonough and Lilien, 1977).

Only those nucleotides which inhibit the accumulation of capping inhibitory activity in organ culture-conditioned media (UDP for retina GDP for cerebral lobes) inhibit the accumulation of agglutination mediating activity. Organ culture-conditioned media prepared in the presence of these nucleotides do not mediate agglutination nor do they compete for agglutinin when added to the fully reconstituted system (Rutz and Lilien, unpublished).

Additionally, in pursuing the purification of the active species in neural retina organ culture-conditioned media, we have been unable to separate capping inhibitory and agglutination mediating activity. Taken together these data indicate that a single species is responsible for mediating specific adhesions and inhibition of capping.

Since we have implicated glycosyl transferases in the turnover of ligand it is profitable to review the hypothesis originally proposed by Roseman (1970) and Roth *et al.* (1971b) that glycosyl transferases are involved in mediating adhesion (see Shur and Roth, 1975, for a recent review). The original hypothesis envisaged the interaction of an oligosaccharide acceptor on one cell with a transferase on another to form an adhesive bond. The value of this hypothesis was that adhesive specificities could be accounted for by the specificities of known enzyme systems. It was also suggested that such a system would provide a ready mechanism for deadhesion and alteration of adhesive specificities — catalysis of the reaction, yielding an oligosaccharide with a new terminal sugar. While our view of the adhesive mechanism is quite different from the original model, it does incorporate these virtues.

Integrating the ideas discussed her we propose that surface adhesive ligand exists in a terminally unglycosylated form in association with a glycosyl transferase (presumably a β-*N*-acetylgalactosaminyl transferase for retina) within an intracellular pool (Fig. 4.4). As the pool of ligand and transferase is mobilized to the cell surface the polypeptide portion of the ligand is exposed and becomes free to interact with the tissue-specific agglutinin. Release of ligand is accomplished by its terminal glycosylation followed by conversion of the transferase to a form recognizing fully glycosylated ligand. The fully glycosylated ligand in turn may interact with cells, restricting both pool mobilization and glycosylation and release. Such a system of feedback controls is absolutely essential if stable adhesions are to be maintained. In the absence of such controls glycosylation and release would continually add free ligand to the intercellular space destabilizing adhesive bonds. Such a lack of control may well be a determining factor in neoplasia and metastasis.

4.6 CONCLUSION

The studies reviewed here address only one aspect of morphogenesis, specific cell adhesion. Studies on cell motility are also clearly central to our understanding of the problem. The close parallels between the dynamics of cytoskeletal and membrane components during cell motility and during the redistribution of surface receptors into caps suggest a common mechanism (see de Petris and Raff, 1973; Lilien *et al.*, 1977). This being the case, inhibition of cap formation by adhesive ligands may reflect their role in controlling cell motility.

REFERENCES

Bahl, O. and Agrawal, (1972), In: *Methods in Enzymology,* Vol. XXVIII. Complex Carbohydrates, (Ginsberg, V., ed.), Academic Press, New York.

Ball, W.D. (1966), The aggregation of dissociated embryonic chick cells at $3°C$. *Nature,* **210,** 1075–1077.

Balsamo, J. and Lilien, J. (1974a), Embryonic cell aggregation: kinetics and specificity of binding of enhancing factors. *Proc. natn. Acad. Sci. U.S.A.,* **71,** 727–731.

Balsamo, J. and Lilien, J. (1974b), Functional identification of three components which mediate tissue-type specific adhesion. *Nature,* **251,** 533–534.

Balsamo, J. and Lilien, J. (1975), The binding of tissue-specific adhesive molecules to the cell surface. A molecular basis for specificity. *Biochemstry,* **14,** 167–171.

Balsamo, J., McDonough, J. and Lilien, J. (1976), Retinal-tectal connections in the embryonic chick: evidence for regionally specific cell surface components which mimic the pattern of innervation. *Dev. Biol.,* **49,** 338–346.

Barbera, A.J. (1975), Adhesive recognition between developing retinal cells and optic tecta of the chick embryo. *Dev. Biol.,* **46,** 167–191.

Barbera, A.J., Marchase, R.B. and Roth, S. (1973), Adhesive recognition and retinotectal specificity. *Proc. natn. Acad. Sci., U.S.A.,* **70,** 2482–2486.

Butcher, F.R. and Goldman, R.H. (1974), Effect of cytochalasin B. and colchicine on α-amylase release from rat parotid tissue slices. Dependence of the effect on N^6, $O^{2'}$-dibutyryl adenosine $3'$, $5'$-cyclic monophosphate concentration. *J. Cell Biol.,* **60,** 519–523.

Cauldwell, C.B., Henkart, P. and Humphreys, T. (1973), Physical properties of sponge aggregation factor. A unique proteoglycan complex. *Biochemistry,* **12,** 3051–3055.

de Petris, S. and Raff, M.C. (1973), Fluidity of the plasma membrane and its implication for cell movement. In: *Locomotion of Tissue Cells.* Ciba Foundation Symposium 14. pp. 27–40. Elsevier, New York.

Edelman, G.M., Yahara, I. and Wang, J.L. (1973), Receptor mobility and receptor–cytoplasmic interaction in lymphocytes. *Proc. natn. Acad. Sci., U.S.A.,* **70,** 1442–1446.

Edwards, J.G. and Campbell, J.A. (1971), The aggregation of BHK 21 cells. *J. Cell Sci.,* **8,** 53–72.

Edwards, J.G., Campbell, J.A., Robson, R.T. and Vicker, M.G. (1975), Trypsinized BHK 21 cells aggregate in the presence of metabolic inhibitors and in the absence of divalent cations. *J. Cell Sci.,* **19,** 653–667.

Garber, B. and Moscona, A.A. (1972), Reconstruction of brain tissue from cell suspensions. II. Specific enhancement of aggregation of embryonic cerebral cells by supernatant from homologous cell cultures. *Dev. Biol.,* **27,** 235–243.

Gottlieb, E., Merrell, R. and Glaser, L. (1974), Temporal changes in embryonal cell surface recognition. *Proc. natn. Acad. Sci., U.S.A.,* **71,** 1800–1802.

Hay, E. and Revel, J-P. (1969), *Fine Structure of the Developing Cornea.* Karger, Basel.

Hausman, R.E., Knapp, L.W. and Moscona, A.A. (1976), Preparation of tissue-specific cell aggregating factors from embryonic neural tissues. *J. exp. Zool.,* **198**, 417–422.

Hausman, R.E. and Moscona, A.A. (1975), Purification and characterization of the retina-specific cell aggregating factor. *Proc. natn. Acad. Sci., U.S.A.,* **72**, 916–920.

Hausman, R.E. and Moscona, A.A. (1976), Isolation of retina-specific cell aggregating factor from membranes of embryonic neural retina tissue. *Proc. natn. Acad. Sci., U.S.A.,* **73**, 3594–3598.

Hautrey A.D., Scott-Burden, T. and Robertson, G. (1974), Inhibition of glycoprotein and glycolipid synthesis in hamster embryo cells by cytosine arabinoside and hydroxyurea. *Nature,* **252**, 58–60.

Henkart, P., Humphreys, S. and Humphreys, T. (1973), Characterization of sponge aggregation factor. A unique proteoglycan. *Biochemistry,* **12**, 3045–3050.

Holtfreter, J. (1939), Gewebeaffinitat, ein Mittel der embryonalen Formildung. *Arch. exp. Zellforsch.* **23**, 169–209, Translated in: *Foundations of Experimental Embryology,* (1964), (Willier, B.H. and Oppenheimer, J.M., eds.), Prentice Hall, New Jersey.

Holtfreter, J. (1943), Properties and function of the surface coat in amphibian embryos. *J. exp. Zool.,* **93**, 251–323.

Humphreys, T. (1963), Chemical dissolution and *in vitro* reconstruction of sponge cell adhesion. I. Isolation and Functional demonstration of the components involved. *Dev. Biol.,* **8**, 27–48.

Humphreys, T. (1967), The cell surface and specific cell aggregation. In: *The Specificity of Cell Surfaces,* (Davis, B.D. and Warren, L., eds.), Prentice-Hall, New Jersey.

Lilien, J. (1968), Specific enhancement of cell aggregation *in vitro. Dev. Biol.,* **17**, 657–678.

Lilien, J. (1969), Toward a molecular explanation for specific cell adhesion. In: *Current Topics in Developmental Biology,* (Montroy, A. and Moscona, A.A., eds.), Vol. 4, Academic Press, New York.

Lilien, J., Balsamo, J., McDonough, J., Hermolin, J., Cook, J. and Rutz, R. (1977), Adhesive specificity among embryonic cells. In: *Surfaces of Normal and Malignant Cells,* (Hynes, R. ed.), Wiley International, England. (In press).

Lilien, J. and Moscona, A.A. (1967), Cell aggregation: its enhancement by a supernatant from culture of homologous cells. *Science,* **157**, 70–72.

Lilien, J. and Rutz, R. (1977), A multicomponent model for specific cell adhesion. In: *Cell and Tissue Interactions,* Soc. of General Physiologists Symposium, (Lash, J. and Burger, M., eds.), (In press).

Marchase, R.B., Vosbeck, K. and Roth, S. (1976), Intercellular Adhesive Specificity. *Biochim. biophys. Acta,* **457**, 385–416.

McClay, D.R. and Baker, S.R. (1975), A kinetic study of embryonic cell adhesion. *Dev. Biol.,* **43**, 109–122.

McDonough, J. and Lilien, J. (1975a), Inhibition of cell surface receptor mobility by factors which mediate specific cell–cell interactions. *Nature,* **256**, 216–217.

McDonough, J. and Lilien, J. (1975b), Spontaneous and lectin-induced redistribution of cell surface receptors of embryonic chick neural retina cells. *J. Cell Sci.*, **19**, 347–368.

McDonough, J. and Lilien, J. (1977), The turnover of a tissue specific cell surface ligand which inhibits lectin induced capping. *J. Supramol. Struct.* (In press).

McDonough, J., Rutz, R. and Lilien, J. (1977), An intracellular pool of a cell-surface ligand which inhibits lectin-induced capping. *J. Cell Sci.*, (In press).

McGuire, E.J. and Burdick, C.L. (1976), Intercellular adhesive selectivity. I. An improved assay for the measurement of embryonic chick intercellular adhesion. *J. Cell Biol.*, **68**, 80–89.

McQuiddy, P. and Lilien, J. (1971), Sialic acid and cell aggregation. *J. Cell Sci.*, **9**, 823–833.

Merrell, R. and Glaser, L. (1973), Specific recognition of plasma membranes by embryonic cells. *Proc. natn. Acad. Sci. U.S.A.*, **70**, 2794–2798.

Merrell, R., Gottlieb, D.I. and Glaser, L. (1975), Embryonal cell surface recognition: extraction of an active plasma membrane component. *J. biol. Chem.*, **250**, 5655–5669.

Merrell, R., Gottlieb, D.I. and Glaser, L. (1976), Membranes as a tool for the study of cell surface recognition. In: *Neuronal Recognition*. (Barondes, S.H., ed.), Plenum Press, New York.

Meyer, J.T. and Oppenheimer, J.T. (1976), The multicomponent nature of teratoma adhesion factor. *Exp. Cell Res.*, **102**, 359–364.

Morris, J.E. (1976), Cell aggregation rate versus aggregate size. *Dev. Biol.*, **54**, 288–296.

Moscona, A.A. (1962), Analysis of cell recombination in experimental synthesis of tissues *in vitro. J. Cell comp. Physiol.*, (Suppl. 1) **60**, 65–80.

Moscona, A.A. (1965), Recombination of dissociated cells and the development of cell aggregates. In: *Cells and Tissues in Culture*. (Willmer, E.N., ed.), Vol. 1, Academic Press, New York.

Moscona, M.H. and Moscona, A.A. (1965), Inhibition of cell aggregation *in vitro* by puromycin. *Exp. Cell Res.*, **41**, 703–706.

Oppenheimer, S.B. (1975), Functional involvement of specific carbohydrate in teratoma cell adhesion factor. *Exp. Cell Res.*, **92**, 122–216.

Oppenheimer, S.B. and Humphreys, T. (1971), Isolation of specific macromolecules required for one adhesion of mouse teratoma cells. *Nature*, **232**, 125–126.

Orr, C.W. and Roseman, S. (1969), Intercellular adhesion. I. A quantitative assay for measuring the rate of adhesion. *J. Memb. Biol.*, **1**, 109–124.

Redman, C.M., Banerjee, D., Howell, K. and Palade, E. (1975), Colchicine inhibition of plasma protein release from rat hepatocytes. *J. Cell Biol.*, **66**, 42–59.

Roseman, S. (1970), The synthesis of complex carbohydrates by multiglycosyl transferase systems and their potential role in intercellular adhesion. *Chem. Phys. Lipids*, **5**, 270–297.

Roth, S. (1968), Studies on Intercellular adhesive selectivity. Dev. Biol., **18**, 602–631.

Roth, S., McGuire, E.J. and Roseman, S. (1971a), An assay for intercellular adhesive specificity. J. Cell Biol., **51**, 525–535.

Roth, S., McGuire, E.J. and Roseman, S. (1971b), Evidence for cell surface glycosyl transferases. *J. Cell Biol.*, **51**, 536–547.

Roth, W. and Weston, J.A. (1967), The measurements of intercellular adhesion. *Proc. natn. Acad. Sci., U.S.A.,* **58**, 974–980.

Rutishauser, U., Thiery, J-P. Brakenbury, R., Sela, B-A. and Edelman, G.M. (1976), Mechanisms of adhesion among cells from neural tissues of the chick. *Proc. natn. Acad. Sci., U.S.A.,* **73**, 577–581.

Rutishauser, U., Yang, C.H., Thiery, J-P., and Edelman, G.M. (1975), Expression and behavior of cell surface receptors in development. Fourth ICN-UCLA Winter Conferences on Molecular and Cellular Biology, (Fox, F., ed.,), Academic Press New York.

Shur, B.D. and Roth, S. (1975), Cell surface glycosyl transferases. *Biochim. Biophys. Acta,* **415**, 473–510.

Steinberg, M. (1964), The problem of adhesive selectivity in cellular interactions: In: *Cellular Membranes in Development,* (Locke, M., ed.,), Academic Press, New York.

Townes, P.L. and Holtfreter, J. (1955), Directed movements and selective adhesion of embryonic amphibian cells. *J. exp. Zool.,* **128**, 53–120.

Trinkaus, J.P. (1963), The cellular basis of fundulus epiboly. Adhesiveness of blastula and gastrula cells in culture. *Dev. Biol.,* **7**, 513–532.

Trinhaus, J.P. and Lentz, J.P. (1964), Direct observation of type-specific segregation in mixed cell aggregates. *Dev. Biol.,* **9**, 115–136.

Turner, R.S. and Burger, M.M. (1973), Involvement of a carbohydrate group in the active site for surface guided reassociation of animal cells. *Nature,* **244**, 509–510.

Umbreit, J. and Roseman, S. (1975), A requirement for reversible binding between aggregating embryonic cells before stable adhesion. *J. biol. Chem.,* **250**, 9360–9368.

Urishihara, H., Takeichi, M., Hakura, A. and Okada, T.A. (1976), Different cation requirements for aggregation of BHK cells and their transformed derivatives. *J. Cell Sci.,* **22**, 685–695.

Weinbaum, G. and Burger, M.M. (1973), Two component system for surface guided reassociation of animal cells. *Nature,* **244**, 510–512.

Vicker, M.G. (1976), BHK 21 fibrobalst aggregation inhibited by glycopeptide from the cell surface. *J. Cell Sci.,* **21**, 161–173.

Vicker, M.G. and Edwards, J.G. (1972), The effect of neuraminidase on the aggregation of BHK 21 Cells and BHK 21 Cells transformed by Polyoma virus. *J. Cell Sci.,* **10**, 759–768.

Yahara, I. and Edelman, G.M. (1972), Restriction of the mobility of lymphocyte immunoglobulin receptors by Concanavalin A. *Proc. natn. Acad. Sci., U.S.A.,* **69**, 608–612.

5 Cell Positioning

A. S. G. CURTIS

Specificity of Embryological Interactions
(*Receptors and Recognition,* Series B, Volume 4)
Edited by D.R. Garrod
Published in 1978 by Chapman and Hall, 11 New Fetter Lane, London EC4P 4EE
© Chapman and Hall

INTRODUCTION

The term cell positioning refers to the processes which determine the placing of one cell type with respect to another. The term 'cell recognition' is often used loosely for these phenomena but it is rather objectionable partly because of its anthropomorphism and also because it refers to other phenomena as well, such as the capture of degraded serum proteins by certain liver cell types.

Cell positioning occurs whenever the arrangements of two cell types or one cell type and some non-cellular class of objects is other than random. Some examples of the main types of cell positioning are given in Table 5.1.

Table 5.1

(a) With a single tissue type
    i.e. positioning cells in various geometric arrays, e.g. around tubules
    e.g. Morphogenesis of liver or kidney tissue
    e.g. Positioning of parts of cells, for instance connections in nervous systems

(b) Between two or more tissue types
    e.g. Embryonic positioning of cell, for instance heart with respect to endoderm
    e.g. Sorting out in aggregates
    e.g. Neuromuscular junctions

(c) Between allogenic individuals
    e.g. Non-coalescence in sponges
    e.g. Non-fusion in *Botryllus* and *Hydractinia*
    e.g. Graft rejection in Gorgonians
    e.g. Possibly graft rejection in vertebrates and other animals

(d) Between different species
    e.g. Specific siting of parasites and some commensals
    e.g. Non-coalescence of unlike species in sponges.

This immediately raises the question of the definition of randomness and its detection in cellular situations. It is clear that quite a number of biologists have misunderstood the nature of random distributions and supposed that the random packings of two cell types would lead to one cell type being evenly distributed in another. Such a distribution is in fact an overdispersed one. A variety of tests for the randomness of distributions exist, e.g. Roach (1968), Pielou (1960), Elton and Tickle (1971). Such tests have very rarely been applied to testing whether cell types are randomly distributed. Obviously there is no great difficulty in stating that some distributions are highly non-random when one cell type clearly separates in some definite topological relationship to the other type. On the other hand it is decidedly difficult to detect small departures from randomness, yet in many experimental situations this is the only hint we shall get of any patterning effect. It is also of interest to detect the level of errors in the patterning mechanism — if some patterns contain 10% of the cells in the wrong place a much less precise mechanism is in operation than if only 0.001% of cells reach the wrong position. Information on this point might be useful in testing the various hypotheses about sorting out.

However, it is not merely sufficient that the positioning of cells is non-random in most biological systems. One cell type must have a definite topological relationship to the other cell type. For example the kidneys have a definite topological set of relationships to a range of other organs in the body. This is simply the general anatomy of the body of that particular species. In other cell systems such as the lymphoid system or the leucocyte or reticulo-endothelial systems positioning is transient for each individual cell though there will be a pattern of distribution of the various member cells of each system.

The development of the nervous system presents a range of phenomena in which there is apparently positioning, not of the whole cell but of the axon extension from the cell body. Here the direction of extension and the final or transient position of functional and non-functional connections presumably parallel the phenomena we see in other cells where the whole cell body takes up new positions. It should of course be remembered that there is considerable whole cell movement in the development of the sympathetic and parasympathetic systems.

The following general types of mechanism can be envisaged which would achieve a positioning of cells of one type with respect to others.

(1) Differentiation of specific types *in situ* in specific positions without cell movement. Cell interactions would determine the types of differentiation and the patterning required. This theory has been, perhaps misleadingly, termed 'positional information' (Wolpert, 1969).
(2) Selective survival of certain cell types in only certain places in the organism.
(3) Migration of the cells, either in the body fluids or by cell movement over cellular or non-cellular surfaces. The movement would be either: (a) Directional, being determined by either negative or positive chemotaxis or kinesis, or by oriented structures determining cell migration. (b) Random, but with trapping mechanism operating to stop cell movement when the cells reached a particular site.

Trapping mechanisms might be an increase in cell adhesion between cell and target so that the cell stops moving, or a cell interaction paralysing the motor mechanisms of the cell or destroying its orientation so that the cell ceased effective further movement. Another trapping mechanism could be that the cells of the target form some kind of sieve so that the migrating cells are unable to force their way through the interstices.

I shall ignore the first mechanism because it is clearly very far from the field of cell migration and positioning. The second mechanism is unlikely by itself to produce effective positioning though there is a theory, see below, in which random movement followed by trapping at all sites and selective survival achieves positioning. It should be appreciated that in real biological systems there are but two main types of positioning mechanism.

(1) Those systems which start from random arrangements of cells and which presumably develop sorting out mechanisms within themselves. The sorting out found in reaggregates is believed to be of this type.
(2) Systems which use a pre-existing non-random arrangement of the cells to develop the pattern.

For instance do lymphocytes accumulate in the lymph nodes and sort out within the nodes because of a system inherent in the lymphocytes alone or do they interact with the stromal cells already prepositioned there. It should be remembered that there is throughout the development of an embryo a non-random arrangement of cell types. In part, this arose from field systems within the egg (see Raven, 1967) and in part, from the operation of positional differentiation during development.

In a few instances in normal development there is evidence that complete breakdown of positioning occurs before the development of a new positioning, e.g. during the development of certain aestivating fish, (annual fish; Wourms, 1972) the embryo disaggregates and then reaggregates so that it develops from a presumably wholly random prearrangement of cells. This resembles the sorting out phenomena seen in aggregates etc. On the other hand it is important to note that in each system seen in whole animals in which it is possible that prepositioning controls the actual positioning that develops, no one has yet separated the contribution of prepositioning and the contribution of positioning mechanisms inherent in the cells themselves.

It is important to make this distinction because positioning *de novo* and positioning dependent on prepositioning have different requirements. The first requires some system which determines position for instance by a scalar or vectorial measurement of some field property, the second merely requires that when the cells are in the correct place some trapping mechanism, such as specific adhesion, operates to add on cells to the existing pattern.

These are the prime requirements for positioning mechanisms.

Only some of the positioning mechanisms envisaged above have been actually examined in any detail. These are:

(1) Random distribution of cells of one type followed by their survival or appreciable mitosis in selected sites. This presumes that there is some nutritive interaction with surrounding cells in only certain sites.

(2) Sieve trapping (in particular in blood or lymph flow).

(3) Trapping of cells by specific adhesion. The origin of this theory is nowadays unclear but Moscona (1962), Roseman (1970), Roth (1968) and Lilien (1968) have strongly, amongst others, advocated this explanation.

(4) The differential adhesion theory proposed by Steinberg (1963) which is a trapping with a low degree of direction of movement.

(5) The morphogen theory developed by myself (Curtis, 1974) which contains elements of trapping and directed movement.

I shall examine each of these theories in turn.

5.1 RANDOM DISTRIBUTION OF CELLS FOLLOWED BY SELECTIVE SURVIVAL

This theory which has connotations of the parable of the sower has a considerable history. Willis (1952) for example, commented on the 'nutritive' theory of malignant tumour distribution. He pointed out that the non-random distribution of some types of malignant tumour could be explained on the concept that tumour cells settled in all manner of vascular and lymphatic sites in the body but that they only survived to multiply in certain sites where the biochemical surroundings and supply of food were advantageous for the cells. It is surprising that almost no experimental work has been carried out to test this theory of the distribution of secondary tumours. A small amount of relevant work has been reported on the distribution of primordial germ cells in bird embryos. Simon (1957) demonstrated that the primordial germ cells in birds arise in the extra-embryonic endoderm and that they appear in the gonads soon after a vascular route has been established between these areas. Dubois and others, in a long series of papers, summarised in Dubois and Croisille (1970) demonstrated that the gonads apparently exert a chemotactic attraction on secondary gonocytes at a much later stage. Thus it was easy to conclude that a chemotactic attraction probably led to the attraction of the primordial germ cells from the blood into the gonads. Meyer (1964) however examined the initial distribution of these cells in the chick soon after their emigration from the extra-embryonic endoderm. He found that these cells appeared in considerable numbers in a wide variety of tissues of the embryo. Unfortunately we know so little about the hemodynamics of the embryonic chick that it is impossible to say whether each volume of tissue has an equal chance of receiving a primordial germ cell but the presence of large numbers of these cells in incorrect sites suggests that the initial distribution of these cells is random. An apparent specificity in the final distribution is achieved by the death of all or nearly all those primordial germ cells that fail to reach the gonads. Possibly teratomas arise from the occasional surviving misplaced germ cell.

Burdick (1968) injected labelled embryonic chick cells into chick embryos and found that the very few cells recovered up to 1 day later showed no signs of homing:

he considered that selective survival or selective lodging (see below) might explain where the cells went.

5.2 POSITIONING AS A RESULT OF FEATURES OF BLOOD OR LYMPH FLOW

The flow of body fluids between cells transports a variety of cells around the body. The flow streamlines at a particular site may bring cells into collision with the surrounding static cells. If the cell is adhesive this may lead to trapping of the cell at that site. Thus peculiarities of fluid flow at certain sites may lead to the selective accumulation of cells at that site. Similarly certain parts of the flow system, such as the smaller capillaries may be so narrow that cells in transport are unable to pass through the capillary and are in effect filtered out of the system at that point. It has been suggested by Wood (1964), Zeidman (1961) and others that the selective positioning of some secondary tumours is due to filtering processes such as these. The pulmonary capillaries are particularly narrow which may account for the positioning of many secondary tumours there.

This field might be well worth further investigation. Though the flow of cells in the blood tends to be an axial one, there may be places where turbulence drives cells against the endothelium. If the force of impingement and the lifetime of the inter-action between cell and endothelium are suitable this might lead to the development of an adhesion between the two. Possibly this may explain the results of Ford *et al.* (1976) on lymphocyte trapping (see Section 5.5.6(b)).

5.3 SPECIFIC ADHESION

In essence this theory proposes that there are a variety of mechanisms of cell adhesion such that two like cells will show some specificity in their adhesion when compared with the adhesions formed between this first type of cell and any other cell type. It should be carefully distinguished from those theories (see later) that propose that the mechanisms that control the adhesiveness of the cell may be in some degree specific. Evidence for species specific adhesion has been recently reported for sponges by McClay (1971, 1974), Humphreys (1963), Muller and Zahn (1973) and van de Vyver (1975); for echinoderms by McClay and Hausman (1976) and for tissue specificity in vertebrates by many authors (see later). Burdick (1970) and Burdick and Steinberg (1969) have reported on the question of species specificity in vertebrates, see also Curtis (1972).

Specific adhesion hypotheses propose one or other mechanisms whereby the actual molecules involved in the adhesion interact in some specific manner.

Early theories, such as those of Tyler (1946) and Weiss (1947) suggested that the adhesion was the result of the interaction of complementary molecules on the surfaces of opposing cells. Later Moscona (1962) and Lilien (1968, 1969) proposed that there were two separate components to the system, an adhesive grouping on the plasmalemmae common to all cells of one type and a material secreted by the cells which interacted with the plasmalemmal grouping. A more recent theory of this type is that of Weinbaum and Burger (1973). Weinbaum and Burger found that the sponge *Microciona prolifera* released a large molecular weight factor on disaggregation (Humphreys (1963) has described the same factor earlier). This factor would allow specifically the formation of large aggregates of sponges at low temperatures. Cells treated with very hypotonic media would not aggregate in the presence of the aggregation factor, thus suggesting that the system had at least two components. They suggested that the hypotonic treatment had released a receptor for the aggregation factor. They termed the material released by hypotonic treatment, 'Baseplate'. If the baseplate was added to hypotonically treated cells followed by the aggregation factor then adhesion occurred. Addition of the components in the reverse order did not produce aggregation. One of the possible interpretations that is compatible with this phenomenon is that the baseplate is attached to the cell surface and that the aggregation factor binds one baseplate to another. Weinbaum and Burger (and Turner and Burger 1973) suggested that the aggregation factor is monovalent for baseplate and that one molecule binds to that from another cell surface by calcium bridging but no precise evidence for this was obtained other than that calcium is required for the system to operate. Another theory of this class involving surface glycosyltransferases was proposed by Roseman (1970) and Roth *et al.* (1971).

It is worth considering the tests and results that would establish such theories. After a factor that produces specific adhesion has been obtained from cells:

(1) The species of molecule that appear to be responsible for specificity should be isolated and purified.

(2) It should be demonstrated that the types of molecules involved are cell surface-associated.

(3) Since it is easy to dissociate cells it is presumably easy to remove one or both components from cell surfaces by such procedures as trypsinisation or treatment with a chelating agent. Consequently, addition of one or both components to a denuded cell surface should re-establish adhesion. In this experiment it is important to demonstrate that the amount of component returned to the surface is comparable to that present in life and that this results in the same type of adhesion of the same strength as that found in life. The reason for this is that if very much larger concentrations are required to produce cell adhesion than act in life, the type of adhesion formed may be quite unlike that normally found in these cells.

(4) The dose—response curve should be examined. It should show evidence for surface binding, and it might be expected that the curve would show a maximum at the level where there is one linking molecule for each binding site if the system requires a bivalent intermediate linking molecule. A univalent form of the linking molecule should block adhesion.

(5) It should also be shown as far as is possible that the molecular species under investigation do not act on cell adhesion by some indirect mechanism such as controlling the arrangment of microfilaments.

Very many studies have shown that a macromolecular factor or factors can be extracted, either from the media that cells have been cultured in or from the media in which cells have been dispersed, which appears to specifically aid the adhesion of cells. Unfortunately very few of such studies have examined the immediate kinetics of cell adhesion which would be expected to be specifically stimulated by such agents.

Macromolecular factors specific to one cell type and increasing the diameter of aggregates or increasing adhesion as assessed by other tests have been reported as follows:

In sponges. Humphreys (1963), Moscona (1968), Weinbaum and Burger (1973), Turner and Burger (1973) for the sponge *Microciona prolifera*; the first two papers demonstrate the specificity of the factor for that species and a similar factor in *Haliclona occulata.* Muller and Zahn (1973), Muller *et al.* (1976a,b) have detected a similar factor in the sponge *Geodia cydonium* which did not stimulate the later stages of aggregation of four *Ircinia* species. These authors suggest that these factors act not on the initial adhesion of cells but on their later aggregation into large aggregates. These factors were obtained by dispersing the sponges in calcium- and magnesium-free sea-water.

In vertebrates. Lilien and Moscona (1967) grew neural retinal cells in serum-free media for 72 hours, a procedure which might well kill the cells. The medium harvested from these cultures enhanced the aggregate size of a variety of cell types but after dialysis the remaining activity was directed specifically to neural retinal cells alone. Antibody against the crude factor (Lilien, 1968) caused the agglutination of whole cells which suggests that cell surface components are present in the crude factor. The dialysed factor increased antibody-induced agglutination but this is difficult to interpret.

Balsamo and Lilien (1974a, 1975) attempted to measure the binding of similar aggregation factors by following the radioactivity after they had been labelled and then added to cell suspension. Binding of homologous and heterologous factors was found provided freshly trypsinised cells were used but the methods used do not show whether the binding was cell surface-located or not. Serum displaced heterologous factors from their binding sites. Balsamo and Lilien (1974b) showed that cells which had bound a homologous aggregation factor and which had then been glutaraldehyde-fixed, though not self-adhesive, would stimulate aggregation of live cells, provided that the live cells could undertake protein synthesis. They suggested on this rather incomplete evidence that the results indicated that not only was aggregation factor required but that another component synthesised by the cells was needed for adhesion. They also showed that growing monolayers release a substance that aids agglutination of fixed cells which had bound aggregation factor. They suggest that these results are compatible with a concept that the adhesion required a cell surface receptor, an aggregation factor and a third component, analogous with the system proposed by Weinbaum and Burger (1973) for sponges. But although the results are compatible with the theory, the cell surface location of any of these components has yet to be proved in order to demonstrate the system. Balsamo and Lilien (1975) demonstrated binding of these factors by cells, but whether to a cell surface location or elsewhere was not detected. McClay and Moscona (1974) obtained similar results. McGuire and Burdick (1976) produced evidence they thought could be interpreted in favour of this theory but which is compatible with many theories of adhesion.

Pessac and Defendi (1972) reported that a variety of cell lines (either murine or human), when cultured in serum-free media produce aggregation factors that enhance the aggregation of various other lines. One line, P388, did not aggregate by itself but could be induced to reaggregate in the presence of conditioned media from several other lines. Some lines (all apparently lymphoblastic) apparently do not produce aggregation-promoting factors. These results are compatible with the hypothesis that some cell types have both receptors for the factor and have the ability to produce the factor, others lack both the ability to respond to the factor and the ability to make it, whilst others can only make factor or only respond. Though there is not very marked specificity in these results it is clear that the system can lead to some degree of specificity. Many workers, e.g. Rutishauser *et al.* (1976) have reported production of non-specific factors by cells promoting adhesion.

Roth (1968) reported that various conditioned media increased the collection of embryonic chick cells of one type by aggregates of the same type in a fairly specific manner.

In other systems. Kondo and Sakai (1971) dissociated echinoderm embryos in 1M urea and EDTA. After embryos had been dissociated in such a solution it contained a factor that enlarged the size of aggregates of homologous species type. No test was made of its effects on the aggregation of other species types. The factor appeared to be identical to a component of the hyaline layer of the egg. Rosen *et al.* (1974) have provided good evidence, meeting most requirements, that a protein located on the cell surface and binding carbohydrates may be involved in genus-specific adhesion in the slime mould *Polysphondylium*.

A fairly large amount of work has been carried out on test systems in which the cells adhere either to a non-living substrate or to a cellular substrate. The latter system was introduced by Walther *et al.* (1973). In the former case the only specificity that can be easily tested for is the specific stimulation (or decrease) of an adhesiveness which may or may not be in itself specific. Unfortunately both variants of the system may be complicated by the involvement of cell spreading in the establishment of the adhesion between the settling cell and the substrate. Takahashi and Okada (1971) report that the adhesion or/and spreading of chick embryonic cells on a plastic surface was specifically stimulated by a homologous conditioned medium.

A further difficulty with the concept of specific adhesion is that it provides no explanation of the patterning of sorting out seen when this develops from random arrangements of cells (Curtis, 1962; Steinberg, 1962a,b,c, 1963, 1970, 1975). Steinberg made the important observations that (a) there is a constant pattern where one cell type surrounds or nearly surrounds a second cell type and (b) that there is a hierarchy of sorting out such that cells lower on the hierarchy always sort out internal to any cell type higher on the hierarchy. These important discoveries point out that we must explain this patterning and that whatever is responsible for the patterning is apparently a graded property common to all the cells he has tested. Steinberg's development of this theory is described and discussed in the next section. Thus, even if there is specific adhesion, it is not by itself sufficient to direct cells into any pattern, while the phenomena we are trying to explain are above all ones of patterning. It might explain the trapping of cells in and around a site where cells of the same or a reactive type have already been localised by other mechanisms. Thus specific adhesion might explain some of the accretion patterns seen in the animal body if it were clearly known that other mechanisms had pre-patterned the trapping cells in a certain place. On the other hand any mechanism that can patterns cells from a random mixture can be used to enhance and add on cells to a pre-existing pattern. It should be noted that there are a number of phenomena, for instance mating types in *Hansenula wingeii* (Crandall and Brock, 1968) and *Chlamydomonas* (Snell, 1976) which almost certainly use specific adhesion to achieve the mating of the correct cell types. Specific adhesion by itself possesses no extended field polarised with reference to the surrounding environment which is essential if patterns are to be formed.

5.4 THE DIFFERENTIAL ADHESION THEORY

Steinberg (1962a,b,c, 1970) explained his findings of a graded quantitative property controlling sorting-out patterns, by proposing that:

(1) The graded property is a range of values of adhesiveness.
(2) The directionality, which places layers of one cell type within the other, arises directly from the relative values of adhesion of the two cell types, both with themselves and with each other.

He suggested that the adhesive mechanism which drives the relocation of the cells (from their initial random to their final sorted-out pattern) is either identical with, or closely analogous to, the mechanism which drives the relocation of two oil droplets embedded in a third medium to their final equilibrium position. This mechanism is described by the surface energy relationships of the droplets with each other and the surrounding medium. Of course such an energetic description is not in itself an explanation. The process is better described by considering the forces involved in the process. One droplet spreads over the other because the adhesion per unit area between the two unlike droplets is greater than the adhesion of the outer droplet surface with the surrounding medium. As the interdroplet adhesion spreads both surfaces of the outer droplet extend and this extension against surface tension tends to act against spread of interdroplet adhesion. Surface tension effects can be expressed as surface energies.

The question is whether this explanation actually operates in biological systems. There are a number of subquestions:

(1) Can a multicomponent system be modelled by a two droplet one ? Antonelli *et al.* (1973), using a rather formal computer approach, estimated that a multi-component system does not sort out appreciably if the surface energy rules are followed. They also questioned whether the approach used by Goel *et al.* (1970, Goel and Leith, 1970) considering the same problem, was correct.
(2) Do surface energy relationships alone operate in cell movement ? It seems more likely that internal cell components such as microfilaments play a role in the motion of cells to new sites, see Harris (1976).
(3) Can cell adhesion be modelled by surface energy relationships ? Such a treatment assumes that the cells come into molecular contact over large areas and that there are no specialised cell contacts.

Steinberg and his associates argue that the observations so far made on sorting out are compatible with their theory. They claimed to show that when aggregates were centrifuged so that they deformed, the increase in surface area that resulted was entirely reversible and entirely an expression of the surface work required to extend the surface. However if the work of pulling a formerly internal cell into the surface, to extend the surface, was in part work against the internal viscosity of the cell the concept that cell position rearrangement is entirely due to surface energy relationships cannot be sustained.

Thus, though the theory of sorting out by the action of differential adhesion has not been disproven no clear experimental evidence in its favour has yet emerged. It is also clear that this theory does not in its simple forms explain a whole range of types of morphogenesis where thin tongues of cells of one type move into a mass of another cell type. Nor does it explain of course those types of cell recognition where no morphogenesis is involved, whereas specific adhesion may be an effective explanation in such situations.

5.5 THE MORPHOGEN OR INTERACTION-MODULATION THEORY

5.5.1 Defects in other explanations

I have criticised the specific adhesion theory because it does not account for the directionality of cell positioning that must occur to explain patterning. Directionality is provided by the differential adhesion theory but there is little experimental evidence in favour of this theory.

The almost classical concept of simple cell recognition systems is provided by the hormone: cell surface receptor system. A soluble substance is bound by a receptor. If a concentration gradient can be set up the system may develop into a chemokinetic or chemotactic one. The other main type of recognition system is that in which both signal and receptor are insoluble and borne on particles such as cells. The mating interaction of the yeast *Hansenula wingeii* (Crandall and Brock, 1968) is almost certainly of this type. The specific adhesion theory has presumed that cell to cell adhesion is of this type but the defect with such a presumption is that there is no directionality in the system. The differential adhesion theory presumes that there is no recognition system but that quantitative difference in adhesion determines the directionality. It is of course possible to imagine that differential adhesion operates together with a certain measure of specific adhesion, but this seems a complex theory.

The work on specific adhesion, though lacking the precision that one would wish, does seem to demonstrate the existence of some type of component of a recognition system. Heretofore it has been assumed that the recognition systems are of the totally insoluble type. I now wish to make the novel proposal that the cell—cell recognition systems are of the soluble signal type, and that the work done in the belief that it was supporting the specific adhesion theory was in fact dealing with the other type of system. The great virtue of the soluble signal/cell surface receptor system is that it provides a directional system if some form of response to a concentration gradient is presumed. It should now also be appreciated that concentration gradient systems can be of two types. The familiar type is one in which the response is positive, e.g. up gradient chemotaxis or kinesis. Some five examples of such systems are well known, e.g. chemotaxis in *E. coli.*, *Dictyostelium* and its relatives, chemokinesis in *Amoeba,* chemotaxis in fern sperm and in leucocytes. The less well-known type of reaction is the negative one, e.g. movement down gradient. Again there is the closely related dichotomy of (i) positive recognition by interaction between two cells and (ii) negative recognition in which the system operates between all, or a wide range of unlike cells, so that the like cells interact not because they recognise each other but because they recognise other cell types as bodies not to interact with. It should be appreciated that again, thinking has tended to limit itself to positive chemotaxis and positive recognition. Thus the examination of any system of cell to cell recognition operating in positioning should answer the questions:

(1) How is directionality obtained ?
(2) Is the system a soluble signal/surface receptor one or an insoluble signal one ?
(3) Is the recognition positive for cells that are to interact, or a negative one in which other cell types are recognised ?
(4) If chemotaxis or chemokinesis operate do they operate as positive or negative systems ?

These questions have already been partially examined in relation to the differential adhesion and specific adhesion theories. I shall now examine these points in relation to the morphogen or interaction modulation theory.

5.5.2 Interaction modulation of cells

This theory was foreshadowed in the theoretical work of Edelstein (1970) who first proposed the use of the term 'morphogen' for positioning factors and first proposed with a fairly detailed experimental examination by myself and Gysele van de Vyver (1971) of the non-coalescence phenomenon in *Ephydatia fluviatilis.* I gave (Curtis, 1974) a fuller statement of the theory with partial experimental evidence for its operation in a number of other biological systems.

In essence the hypothesis proposes that most of all cell types within an organism produce soluble diffusible substances (morphogens) that diminish the adhesiveness of unlike cell types. Obviously it cannot be expected that these substances will react with all cell types and it would be likely that cells from very different species would normally be unable to interact in this manner. Concentration gradients of these substances would be set up so that, close to one tissue, other cell types would be non-adhesive. It can easily be seen that if a cell is able to respond differentially over its length to such a gradient chemotaxis will result. Alternatively if the cell can retain the effects of its interaction with a given concentration of the substance for a certain time chemotaxis or chemokinesis may result.

Presumably these substances interact with cell surface components which are either directly or indirectly involved in cell adhesion. There are two possible systems for controlling the specificity of such interactions. First, each substance is specifically prevented from reacting with the cell type that produces it, while the interacting cells bear a common and wide-spread receptor that interacts with the morphogen. Alternatively, the interaction is with a limited range of receptors borne by the interacting cells.

It is worthwhile digressing for a moment to notice that the morphogen theory is but one of a set of similar theories that can be proposed. namely:

(1) Positive interaction of like cells — no effect on unlike cells, leading to positive chemotaxis. This is the concept originally put forward by Townes and Holtfreter (1955) and elaborated by Edelstein (1970).

(2) Positive interaction of unlike cell types.

(3) Negative interaction of unlike cells so that cells of unlike type tend to move down gradient.

(4) Negative interaction of like cells so that they tend to disperse.

It should be noted that two variants of the theory suppose that a cell recognition system operates in which like recognises like while the other pair of variants of the theory propose that cell recognition is of unlike types. It is perhaps rather ironic that heretofore it has often been tacitly assumed by students of cell recognition that it must be a recognition of like while there is clear evidence that the relatively well-understood phenomenon of antigenicity is a recognition of unlike.

Since the classification set out above predicts the existence of four types of morphogen I propose at this point to term those morphogens that act so that unlike recognises unlike and that unlike do not associate together, *interaction modulation factors*.

The third form of the theory is of course that which I proposed (Curtis, 1974, 1976). All these systems could lead to cell positioning. It should be noticed that combination of the first and third forms of the theory might produce a most effective method of cell positioning.

It is worth noting that four types of mutant in any of the four types of system might be expected. These are respectively (i) loss of ability to produce morphogen, (ii) loss of ability to respond to morphogen (iii) production of an aberrant morphogen with resemblances to that of another tissue, (iv) development of an ability to respond to other morphogens. It should be appreciated that there may be one or more cell types in an organism that do not react to any morphogen, and produce none. It seems possible that fibroblasts are one of these types. Fibroblasts are found in very many tissues and of course in connective tissue which joins one tissue to another. There is also reason to suspect that fibroblasts from two different tissues may not show sorting-out behaviour (Tickle, in preparation).

5.5.3 The experimental requirements for the demonstration of an interaction-modulation system

It is well worth considering the tests that must be carried out to demonstrate that interaction-modulation factors operate, for this helps understanding of the system. Initially it must be shown that cells of one type produce a factor that diminishes the adhesiveness of other cell types when this is measured using a cell suspension to examine cell to cell adhesion. Obviously such a factor should be produced by cells which are actively engaged in normal metabolism. The factor should not affect the adhesiveness of the cell type from which it is derived. The factor should not be toxic and any effect it has on cells of a certain type should be reversed after the factor is removed.

Addition of an interaction-modulation factor to a mixture of two types of cell should produce the appearance of specific adhesion because the adhesiveness of one type will be so reduced that the aggregates that form will be composed predominantly of the other cell type, that produced the factor. This experimental result cannot by itself be distinguished from that in which 'ligand' promoted adhesion. However, if two cell types and their respective factors are mixed, a clear test between the two hypotheses is obtained. If the factors do indeed reduce the adhesiveness of cell types heterologous to them, then when two are mixed in the presence of their heterologous cell types the adhesiveness of both types will be diminished. If the 'factor' was in fact a ligand, both cell types would have their adhesiveness for their own type raised.

Even if a system passes all these tests it has not been proven to be an interaction-modulation system of the type described above. The vital test of such systems is that imposition of a non-normal concentration gradient of such a factor in a system will either derange cell positioning or even produce a new abnormal positioning.

Unfortunately, there are very few experimental systems in which an abnormal concentration gradient of factor can be applied to a mass of cells. It will be shown shortly that the sorting out patterns seen in aggregates can be explained on the hypothesis that the medium around the aggregates is a sink for the concentration gradients set up in the aggregates. As a consequence the experimental situation in which a high concentration of an interaction modulation factor is added to the medium around an aggregate is not one which will set up a long-lasting concentration gradient unless the factor is continually destroyed in the inner regions of the aggregates. Thus this type of experiment is one which is not an ambiguous demonstration of the action of such a system. In order to produce long-lasting gradients, artificial sources and sinks must be introduced. Another way of detecting whether these substances act through the existence of concentration gradients is to impose an even concentration field on the system. If concentration gradients are responsible for the sorting out, their abolition should destroy any sorting out.

5.5.4 Evidence for the existence of interaction-modulation systems

Studies on cell adhesion require a rapid and accurate method of measuring cell adhesion. In early work the diameter of aggregates was used as a quantitative measure of cell adhesion. Unfortunately this measure assumes (i) that the final aggregate size is one at which there is an equilibrium between accretion and break up, and (ii) that there have been no problems of nucleation in the formation of the aggregate, in other words that all cells are equally likely to form the first adhesions. This measure also takes many hours to obtain so it represents both the past and present adhesiveness of the cells. The first assumption mentioned above has only been tested on two occasions, see Gerisch (1968), and by Curtis (1973), with differing results for differing material. The second assumption appears never to have been tested.

Curtis and Greaves (1965) introduced the idea of looking at the early kinetics of cell aggregation. This concept was further developed by Curtis (1969) and Curtis and Hocking (1970) to provide a quantitative method measuring adhesion, in which collision efficiency is used as a measure of adhesion. Collision efficiency is the probability that a collision between cell and cell, cell and aggregate, or aggregate and aggregate, results in an adhesion. Following the theoretical treatments provided by Swift and Friedlander (1964) it is clear that such measurements must be made in a known laminar shear flow. Such a flow is easily provided by a Couette viscometer. The main advantage of this technique is that a measurement of cell adhesiveness is obtained which is independent of effects of cell size (so that different cell types can be compared), shear rate in the medium, cell concentration and to a large extent, of medium viscosity. The measurement is made at low rate so that the cells are not appreciably affected by the motion. The interaction time of a collision is so short that the measurement is not appreciably affected by cell spreading though of course this may act to stabilise the adhesions against the effects of much greater shear rates.

This technique can be extended to detect specific cell adhesion. I pointed out (Curtis, 1970a,b) that if mixtures of two cell types are made, the adhesiveness of these mixtures will be a weighted mean of the adhesiveness and proportions of each single type if adhesion is non-specific, but if adhesion is specific the adhesion will be much reduced in mixtures because unlike cells will not stick.

Sponges. Using this technique I was unable (Curtis, 1970a,b) to find any evidence for specific adhesion of the cells of the marine sponges *Halichondria panicea, Microciona sanguinea, Hymeniacidon perleve* and *Haliclona occulata.* The cells used in these studies had been very well washed after disaggregation. van de Vyver and I then examined the adhesive situation between various strains of the freshwater sponge, *Ephydatia fluviatilis* (Curtis and van de Vyver, 1971). Each strain type had previously been shown to be non-coalescent with another strain type (van de Vyver, 1970). At first we found, using well-washed cells, that the mixtures showed no sign of specific adhesion. However when cell suspensions containing some of the medium in which disaggregation had been carried out were used, the adhesiveness of mixtures of two strain types was reduced to a value well below that to be expected if the strain types were entirely specific in their adhesion. This suggested that the strain types might secrete substances that diminish the adhesiveness of the opposite strain type. We tested this and found that the various strain types produce such substances. Cells recover their adhesion after removal of such substances. At this point, the status of aggregation-promoting substances in sponges should be considered. Humphreys (1963), Moscona (1968) and Muller and Zahn (1973) have reported that several species of sponge contain substances that specifically promote species-specific aggregation (see above). When an aggregation-promoting factor of one species type is added to a mixture of cells from its own and a different type, the aggregates formed appear to be only of the same species type as the factor. Is this due to specific promotion of cell adhesion or to the suppression of adhesion of the other cell type ? This matter clearly needs investigation anew.

I then examined graft rejection in the sponge *Hymeniacidon* sp. (Curtis, in press). This sponge shows very marked non-coalescence. Graft rejection between pairs parallels non-coalescence. Wherever a pair show graft rejection or non-coalescence the sponges produce factors that diminish the adhesion of the cells of the other strain type.

Thus it can be seen that the work on sponges has shown with a variety of species that interstrain interactions are at least to a considerable degree due to the action of substances that show most of the properties of interaction-modulation factors. Unfortunately, their role, if any, in interspecies interactions is yet to be examined. However, no experimental work, other than in *Hymeniacidon* (Curtis 1978), has yet been carried out which shows that cell positioning in sponges is affected by such factors, though the non-coalescence phenomenon is observational evidence for such a system. Kondo (1974) produced evidence for the existence of aggregation inhibitors in echinoderm tissues.

Avian tissues (aggregates and related systems). The combination of embryonic chick liver and neural retina cells has recently provided further evidence for the existence and operation of an interaction-modulation system. This system has been much studied and has in turn provided evidence in favour of both the specific adhesion and differential adhesion theories of positioning. Steinberg (1962a,b,c) showed that the two cell types (from 7 day embryos) sort out in aggregates so that the liver segregates internally to the neural retina. Roth and Weston (1967) and Roth (1968) using the collecting aggregate system and Walther *et al.* (1973) the collecting cell lawn technique, showed that neural retina aggregates and lawns collect neural retina cells in preference to liver from mixed suspensions while liver aggregates and lawn show a preference for liver cells. These experiments were carried out under conditions in which the collecting cells might have conditioned the medium. McGuire and Burdick (1976) improved technical details of the collecting aggregate system and were thus able to get a very marked apparent specificity of collection for a number of cell type combinations, (neural retina-liver-mesencephalon). Again the collection was carried out in a conditioned medium.

I re-examined this system because earlier work (Curtis, 1970a,b,c) suggested that no specificity of adhesion could be detected when the kinetics of aggregation of mixed suspensions was examined. However these cells were well-washed and were only examined for 4 hours after disaggregation and it is arguable that the non-specific phase noted as an initial phenomenon by Roth was merely accentuated in my work. The composition by cell type of the first aggregates formed was examined. If there is no specificity of adhesion and the cells are equally adhesive the aggregates formed in a 50:50 mixture of two types should also contain this proportion. If however there is total specificity of adhesion the aggregates will be either one cell type or of the other. Liver and neural retina cells have approximately the same adhesiveness so that they are suitable for this test. Results have been described in detail (Curtis, 1974) but Fig. 5.1 shows representative results. The results are clearly compatible with non-specific adhesion. However it should be remembered that the same appeared to be true for sponge cells when they were well-washed. Consequently,

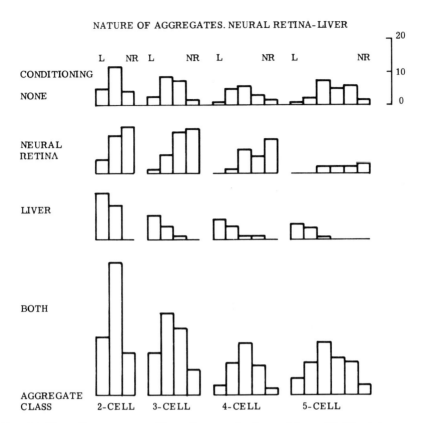

NATURE OF AGGREGATES. NEURAL RETINA-LIVER

Fig. 5.1 The cell-type composition of aggregates formed from 50:50 mixtures
of embryonic chick liver and neural retina as an indication of specificity. Effects
of interaction-modulation factors on the composition. The histograms show
the number of aggregates of given cell-type composition observed for 2, 3, 4
and 5 cell aggregates. Thus an aggregate may be a 4 cell one, which will fall
into one or other of the following classes; pure neural retina, 3 retinal cells
to one liver, 2 retinal cells to 2 liver, etc.

In the absence of any added IMF the cell-type composition of these very
early aggregates formed in the first hour of aggregation is random, i.e. no
specificity, with mixtures being most common. Adding neural retina IMF
biases the aggregate composition towards richness in retinal cells, while liver
IMF biases towards richness in liver cells. This result is not inconsistent with
explanation in terms of specific adhesion but the fourth line of histograms
shows that the composition in the presence of both IMFs is not consistent
with explanation in terms of specific adhesion. The histograms obtained with
different treatments are not to be compared together to indicate that more
or less total adhesion took place because the data was acquired by scoring
the composition, using fluorescent antibody methods, of aggregates until
some 150–200 aggregates had been examined. The number of single cells
encountered in such searches is not recorded and should not be assumed
to be identical for different treatment.

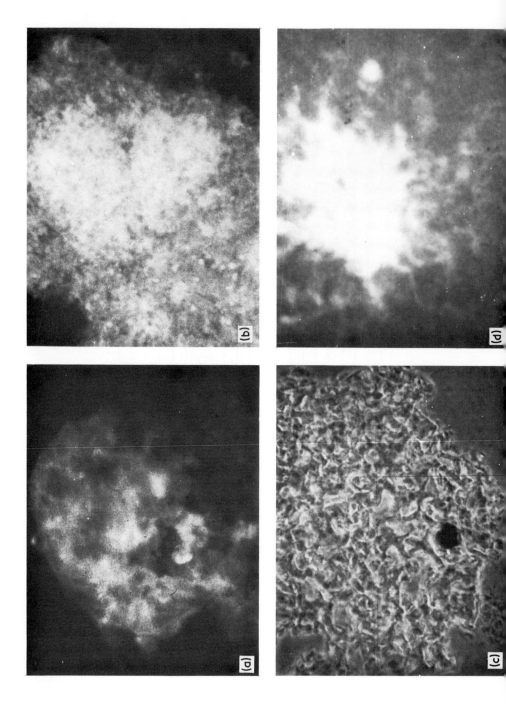

Fig. 5.2 (a and b). Early stages in sorting out in 7 day embryonic neural retina: liver aggregates in absence of added IMFs. Mixed light (a) at 4 hours after start of aggregation, × 580 (b) at 24 hours, when the sorting-out is not complete but internal location of liver is appearing. × 290. (c and d) Sorting out at 43 hours in a control: neural retina: liver. (c) Phase contrast, × 580. (d) Fluorescence. × 580. Note internal location of liver. Frozen sections.

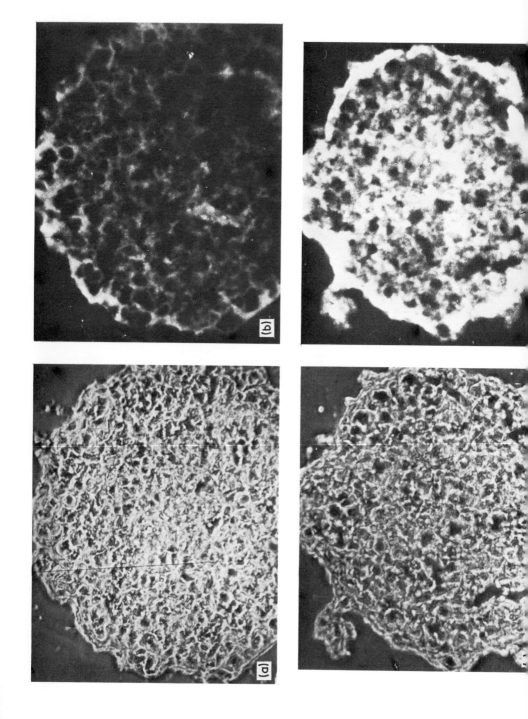

Fig. 5.3 (a and b) Effect of interaction modulation factor on sorting-out of 7 day embryonic heart: liver combination at 48 hours of culture. (a) Phase contrast, (b) Fluorescence. Heart IMF applied from 4 to 48 hours of culture. Note liver has reversed to form a continuous external layer. × 580. (c and d) Reversal of sorting-out in neural retina: liver aggregates caused by treatment with neural retina IMF. Liver is located as the continuous external phase, contrary to the situation in the control, see Fig. 5.2 (d). (c) Phase contrast, (d) Fluorescence. IMFs applied from 4 to 48 hours. In this figure the photographs are not of the same section by two types of illumination but of two successive sections. × 580. Frozen sections.

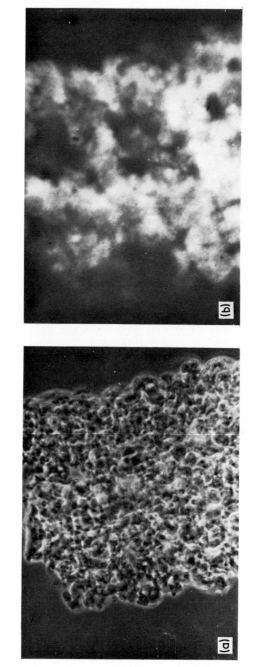

Fig. 5.4 (a and b) Breakdown of sorting-out in neural retina: liver frozen sectioned aggregates caused by liver IMF, applied from 4 to 48 hours. (a) Phase contrast, (b) Fluorescence. × 580.

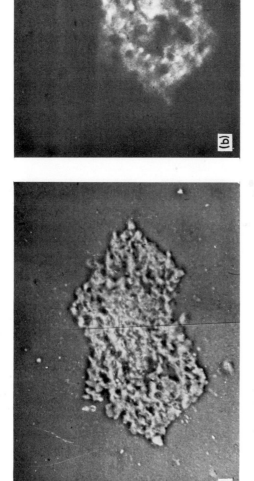

Fig. 5.5 (a and b) Both internal and external sorting-out of liver in a neural retina: liver frozen sectioned aggregates when both liver and retinal IMFs applied from 4 to 48 hours. (a) Phase contrast, (b) Fluorescence. × 580.

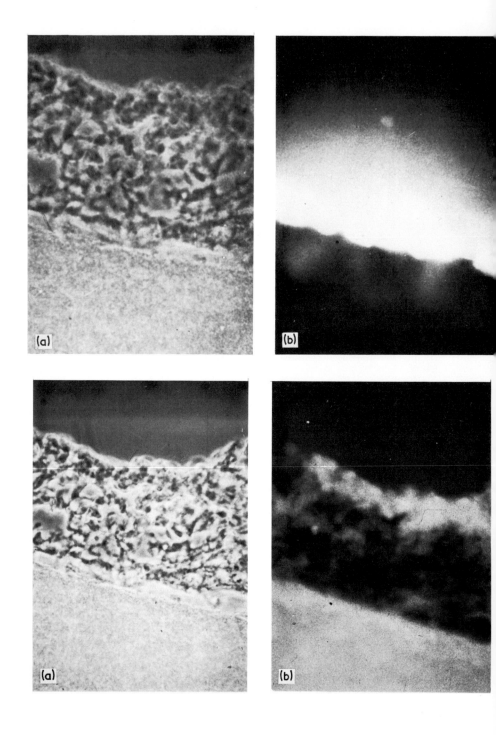

Figs. 5.6 and 5.7 Effects of diffusion gradients of IMFs on sorting-out of neural retina: liver mixtures. Cells grown in thick layers (Ultra-high density cultures) on Pellicon PSJM filters. Frozen sections. Fig. 5.6 Control group with no IMF added. (a) Phase contrast, (b) Fluorescence. Note internal position of liver. × 580.

Fig. 5.7 Group with neural retina IMF permeated through the filter from the 'cell' side of the filter. (a) Phase contrast, (b) Fluorescence. Note reversal of sorting-out × 580.

I repeated these experiments using media which had been conditioned by culture of cells. Mixtures of cells aggregated in medium conditioned by one of the cell types show specific adhesion (Curtis, 1974) because the aggregates show non-random compositions. Do the conditioned media contain a factor that enables cells to stick together specifically or do these factors act in an exactly the opposite manner? Namely, do they reduce the adhesion of unlike cell types? Two tests were carried out on this, first, measurement of the effect of the media on the adhesiveness of a single cell type showed that the adhesiveness of unlike cell types was lowered while that of the homologous type was neither raised or lowered. Second, the composition of aggregates formed from mixtures and containing factors from both cell types is a random one; if the factors specifically increased the adhesion of their own cell type to itself the aggregates would be either of one type or the other, but never mixed.

However, these results though entirely compatible with the concept of the control of cell positioning by interaction-modulation factors do not prove it.

5.5.5 Crucial tests of the interaction-modulation system hypothesis

The crucial test of the action of such systems is to show that the application of abnormal concentration gradients of such factors leads to (i) randomisation, or (ii) abnormal patternings of cell distribution. Obviously *in vivo* experiments are attractive but there may be problems (i) with the maintenance of gradients and (ii) with the possibility that other positioning systems have been disturbed.

To date the system studied by Richard Hoover and me looks promising. Aggregates of binary mixtures of neural retina, heart and liver are formed and the cell distribution studied by using an immunofluorescent technique.* Controls at four hours (Fig. 5.2) show little or no sign of sorting out but considerable signs at 24 hours. By 48 hours, sorting out has taken place and neural retina or heart surround liver. In the first type of experiment either neural retina or liver IMFs are added to the surrounding medium after four hours. If we assume that there is a sink for these factors in or on the cells then a long-lasting concentration gradient will be set up from the outside to the inner regions of the aggregate. If no such sink exists then there will be a short-lived concentration gradient though, because of the probable molecular weights of IMFs and the nature of aggregates the gradient may persist for many hours. Consequently the question we ask is whether abnormal concentration gradients alter sorting out. The results, see Figs. 5.3 and 5.4, are that NR IMF reverses sorting out, so that liver now segregates externally while liver IMF prevents sorting out. Very similar results are obtained if these gradients are applied after 24 hours of sorting out though the reversal of liver position is less clear. It should

---

* Fluorescence microscopy (Figs. 5.2–5.7) was carried out using TRITC tetramethyl rhodamine isothiocyanate) – conjugated goat anti-rabbit IgG-stained sections after reaction with rabbit anti-chik embryonic liver (11 times absorbed). Thus the fluorescence reveals the location of the embryonic liver cells.

be noted that some liver lies internally but that liver is the external continuous phase while neural retina is the internal discontinuous phase in these aggregates. This corresponds to Steinberg's definition of sorting out (Steinberg, 1963). In normal aggregates, liver is the internal discontinuous phase. Finally when both IMFs are applied simultaneously we get aggregates where liver cells sort out both externally and very internally as though two processes are competing (Fig. 5.5).

Thus these experiments provide very strong evidence that there are factors that will alter cell positioning, though of course they give little evidence as to the actual mechanism.

It should also be noted that although it is almost certain that concentration gradients are set up in these experiments and though it is easy to envisage control mechanisms involving concentration gradients there is no direct evidence that concentration gradients are involved.

A further set of experiments suggests much more strongly that concentration gradients are involved. A thick layer of cells is grown on a Millepore Pellicon filter. Sorting out occurs in controls with neural retina external (Fig. 5.6). When medium containing neural retina factor is permeated through the cell layer and filtered from on top at the slow rate of ca. 1 ml/cm^2/day liver takes an external position. When the permeation is from underneath liver retains its bottom position, Figs. 5.6 and 5.7.

Thus this system provides much experimental evidence in favour of the action of such factors in cell positioning.

5.5.6 Evidence from other systems

(a) *Fibroblasts*

Dr. D. Matthopoulos and I have separately examined a number of fibroblasts lines, e.g. CHO, BHK, and L929 to discover whether these cells produce IMFs or respond to them from other tissues. No evidence was found for cross-species reactivity with other cell types when mouse liver or chick neural retina or liver were used, similarly mouse liver did not react with L929 cells. Thus the preliminary evidence suggests that fibroblasts may be a cell type that has escaped from the IMF system. It is of interest of course to recall that fibroblasts form in many ways a tissue that extends to nearly all parts of the body without regard to other tissue type.

(b) *Lymphocytes*

I and Maria de Sousa (1973, 1975) discovered that rat, mouse, and human B lymphocytes produce substances that diminish the adhesion of T lymphocytes and vice-versa. These factors have molecular weights around 1×10^4 and are proteins (Curtis, in preparation). They are bound to target cell surfaces. The T interaction-modulation factor appears to be a suppressor of the pokeweed mitogen response by B cells, (Haston, in preparation) and if injected into syngeneic mice alters the localisation of labelled lymphocytes, directing them particularly away from the lymph nodes to the spleen, (Davies, in preparation). In view of the recent work by Ford *et al.* (1976)

and Sedgley and Ford (1976) which shows that neuraminidase-treated lymphocytes have a different vascular circulation rather than a different specificity of adhesion, any finding which claims that an alteration in cell localisation is a result of surface alterations must be regarded with caution. Nevertheless, the fact that these factors are normal products of the cells involved, binding to the target cell, altering their adhesiveness and changing localisation suggests that they may be the actual control substances for lymphocyte localisation.

Lymphocyte recirculation, otherwise named ecotaxis or homing, provides a series of cellular interactions in which recognition phenomena may be very important. Recognition may occur at the endothelial surface where lymphocyte enters the lymph node. Some degree of sorting out occurs within the node or spleen so that cells are allocated to the well-known T- or B-dependent areas. Possibly some degree of sorting out occurs in the reaction of lymphocytes with other tissues and there is evidence for lymphocyte mis-positioning in certain diseases like mycosis fungoides. The work of Ford and Simmonds (1972) which shows that there is no inherent property of fast or slow circulation through a lymph node for a given cell, suggests that the recognition properties of lymphocytes may not be immutable properties of those cells.

The discovery of the lymphocyte interaction-modulation factors allows re-interpretation of many features of circulation in a more coherent manner. For instance, Sprent (1973) observed that the recirculation of lymphocytes in the B mouse is very slow. Lymph node cells from B mice are abnormally adhesive (Curtis and De Sousa 1975). Thus it can be argued that, in a normal animal, secretion of the T IMF in the lymph node reduces the adhesion of the node cells so that they fall out into the lymph duct much more readily than in a B mouse where T IMFs are missing.

Animals in which the T population has been partially removed by low doses of ALS (antilymphocytic serum) showed enhanced response to antigen (De Sousa and Haston, 1976). This seemingly unlikely result can be easily explained if it is supposed that T cells produce a factor which normally stops T cell co-operation with B cells. Bell and Shand (1975), using rats, found that adding normal thoracic duct lymphocytes to primed ones and then injecting the mixture into lethally irradiated hosts reconstituted with bone-marrow (B) cells led to much smaller antibody responses than in controls with primed lymphocytes alone. The larger the addition of normal lymphocytes the smaller the response. This result is readily explained on the idea that excess T IMF production tends to suppress responses to antigen, presumably blocking T—B co-operation. Bell and Shand (1975) also discovered that attempts to fill unoccupied lymphoid areas in irradiation-depleted rats with lymphocytes led to a considerable increase in recirculating lymphocyte populations. This again is consonant with the idea of the activity of T and B factors in controlling recirculation. Related experiments described by these authors are also interpretable in the same manner.

B and T lymphocytes sort out in the nodes and the spleen, see for example Nieuwenhuis and Ford (1976) in a manner which may be an excellent model for segregation in aggregates.

Finally it should be appreciated that T cell suppression of the immune response could be very simply viewed, in many cases, as an example of the action of T factors preventing T-B co-operation.

(c) *Histocompatibility effects*

The work on graft rejection in sponges (Curtis, 1977) shows that histocompatibility is associated with interaction-modulation factors. Since tissue interactions in higher animals may be controlled by similar factors we can then ask whether histo-incompatible reactions in higher animals are controlled in a similar manner and whether the two systems may not be so closely related that the relationship could be summarised by stating that tissue interactions have evolved into histocompatibility ones or vice-versa. In all previous work on sorting out in aggregates no attention to the question of whether histocompatible or incompatible tissues from animals of different genotype were mixed (though it is highly probably that a certain amount of mixture of potentially histo-incompatible cells took place simply because most species have many histocompatibility types and because no particular care to obtain syngeneic cells was used).

The interaction-modulation factors from lymphocyte populations and from liver diminish the adhesiveness of identical cell types if they are of different histo-compatibility type, Curtis (in preparation), besides reducing the adhesion of different tissues of the same histocompatibility type. This finding suggests that the tissue effects and histocompatibility effects may be very closely linked since the lymphocyte IMFs involved in histocompatible effects are known to be defined very precisely in genetic terms. They are specified by either the D region of the H-2 complex in mice or by a locus which is closer than S to D. When a factor from one tissue is mixed with cells from another tissue of different H-2 (D region) type, it appears that the two effects sum together.

An antibody to a given T factor will in its Fab form inhibit the effect of that T factor on a target cell, and, as complete antibody, will also produce H-2 D region-specific complement lysis of target cells if and only if they have the same constitution at D as did the cells from which the factor used to raise the antibody came. All these observations strongly suggest that the histo-incompatible effects and the tissue effects within an organism's tissues are one and the same.

If this supposition is correct we would expect that cells, say from two livers of different histocompatible types, might sort out in aggregates. Results of appropriate experiments suggest that this is so.

5.5.7 The possible nature and role of interaction-modulation factors

It can be seen that evidence has been found for the action of interaction-modulation factors in sponges, (probably allogeneic combinations), in syngeneic combinations of embryonic chick tissue and in both syngeneic and allogeneic combinations of various mammalian tissues. These factors diminish the adhesion of unlike cell types and in the chick embryonic system there is evidence that cell positioning may be affected as well. This effect of these factors on adhesion, which produces an apparent specificity of adhesion, is compatible with nearly all those experiments which appear to demonstrate specific adhesion. The only experiment which distinguishes between these explanations is to aggregate a mixture of cell types in the presence of factors derived from both cell types and to examine the effect of the factors on the adhesion of unlike types. If the interaction-modulation explanation is correct, the aggregates should become less specific and the cell adhesion weaker. On the other hand, if the specific adhesion theory is correct the aggregates should become more specific and each cell type more adhesive for its own type. The mere stimulation of the adhesion of its own cell type alone by a factor is not proof of the specific adhesion theory because the effect may be a non-specific increase of adhesion. The specificity is tested by trying the factor on unlike cell types.

It is also interesting to note, perhaps from the vantage point of an uninvolved onlooker, that specific adhesion theories suppose that recognition is of like by like. The interaction-modulation theory is one which supposes that recognition is of the unlike. I am reminded of the adage 'Know one's enemies'. It is interesting to note that those recognition systems which are relatively well understood, e.g. the classical immune systems, are those in which unlike is recognised and that no biological system is known in which like by like recognition has been clearly established. However this argument must not be over-relied upon.

The interaction-modulation theory is as stated above, deficient at the moment in evidence regarding positioning. However studies of lymphocyte traffic have revealed many phenomena which are peculiarly effectively explained by the theory.

It is not clear at the moment whether the interaction modulations seen in combinations of allogeneic tissue are of the same type as those seen between different tissues in one syngeneic organism. The phenomena in allogeneic combinations only produce very simple positioning and often nothing more than a drop in the adhesion of one or both tissue types. It is worthwhile speculating whether such systems play a role in graft rejection in vertebrates. Some features of graft rejection suggest that the graft cells lose adhesion (see Medawar, 1944). It is easy to see that genetic differences would ensure that the two allogeneic tissues had unlike interaction-modulation systems. If the interaction modulations of tissues within a syngeneic system act in a parallel manner it must be supposed that there are a range of loci, different sets of which are expressed in each tissue. Alternatively, uniqueness is given to tissues by ensuring that although there are only a few loci involved, a large number of possible combinations of alleles exist, such that many different tissues could have their own unique combinations.

One interesting offshoot of these concepts is the idea that there may be a tissue type which has no ability to interact with others. The consequence is that this type will adhere to other cell types without any possibility of losing adhesion. Such a cell type would of course act to bind different tissues together. Thus the almost ubiquitous fibroblast would be the ideal candidate for the type. It is remarkable that no evidence for interaction-modulation factor production or action (in syngeneic systems) has yet been found for fibroblasts (see above). However, the evidence is not yet based on a wide enough range of fibroblasts and I merely raise it as an interesting area for further research.

Similarly it is easy to imagine that there should be diseases of such a system. The disease might take the form of the loss of the ability to respond to factors from other tissues so that small numbers of such a cell type could easily infiltrate, by cell movement, into a large mass of the normal tissue. Conversely, factor production might stop so that the cells would not produce any reaction in neighbouring small groups of cells and might be infiltrated by them. All such changes would be a type of abnormal differentiation probably classifiable as tumours.

A further speculation is that abnormal differentiation, as in a tumour, might lead to one cell type acquiring the receptors or factors typical of another. As a result the tumour would tend to metastasize to very specific secondary sites.

I have already suggested that there may be a close relationship between the actions of interaction-modulation factors and histocompatibility antigens. The main reasons are (i) because of the involvement of IMFs in allogeneic reduction of adhesion between tissues of unlike H-2 type, (ii) because graft rejection involves a loss of adhesion together with cytotoxic effects, and (iii) because of the genetic similarities between H-2 systems and the allogeneic display of IMF action. Suggestions have been made that the precise expression of histocompatibility antigens varies from tissue to tissue within an organism. This would provide the difference required to ensure that an interaction-modulation system might operate between tissues within an organism.

There are already reports which suggest that genes located in the histocompatibility loci are involved in B-T interactions. For instance Katz and Armerding (1975) reported that such genes in the H-2 complex are involved in the production of factors that aid B-T interactions. Similarly, the MLR and CML (cell-mediated lympholysis) reactions are closely related to differences in the histocompatibility loci. A number of authors, including McDevitt (1975) and Greaves, Owen and Raff (1974), have proposed such a control system for lymphocyte interactions. It is now possible to see that such a genetic system might underlie all cell—cell adhesion reaction.

5.5.8 Models for the interaction-modulation system

All the models I am about to postulate propose that there are receptors of some type for interaction-modulation factors and that effects on adhesion may be direct or indirect.

The first model to be considered suggests that the receptors which bind the IMFs are surface antigens connected with the histocompatibility system. Thus two cell types with identical sets of histocompatibility antigens will bind all the IMFs they produce. As a result there will be no free IMFs to affect cell adhesion. One might propose that the IMFs are proteases. Such a theory of course supposes that IMFs are being continually produced and consequently require that the cell surface is fairly rapidly turned over so that there is a continual supply of new receptors to bind IMFs. This theory has the unsatisfactory feature that a set of very specific proteases have to be coded for which exactly parallel the histocompatibility antigens.

When two cell types meet, unlike either in the types of antigens and IMFs they display, or alternatively in the degree of expression of some of their antigens, one or both cell types lack either the complete set of antigens to inhibit the IMFs produced by the other cell type, or alternatively sufficient antigen to prevent IMF attack. As a consequence their IMFs attack the plasmalemmae of the other cell type and its adhesion drops.

The second model proposes that the histocompatibility antigens are directly or indirectly involved in adhesion and that breakdown products of such antigens shed into the surrounding medium, e.g. microglobulin, are in fact the IMFs. It is necessary to suppose that breakdown fragments of one set of histocompatibility antigens can interact with unlike antigens.

The third model suggests that the IMFs are direct cell-to-cell binding agents. When heterologous combinations of cell and IMF are made, the abnormal IMFs bind to the receptors and block them from action in cell to cell adhesion. This theory has the difficulty that there is little evidence that IMFs are promoters of homologous cell adhesion.

Obviously each of these theories has a good deal of evidence in its favour, described earlier in this chapter. Similarly each theory has several areas of weakness. Investigation of the IMFs is still at too early a stage to accept or reject each or any of these theories.

REFERENCES

Antonelli, P.L., Rogers, T.D. and Willard, M.A. (1973), Geometry and the exchange principle in cell aggregation kinetics. *J. theor. Biol.,* **41**, 1–22.

Balsamo, J. and Lilien, J. (1974a), Embryonic cell aggregation: kinetics and specificity of binding of enhancing factors. *Proc. natn. Acad. Sci. U.S.A.,* **71**, 727–731.

Balsamo, J. and Lilien, J. (1974b), Functional identification of three components which mediate tissue-type specific embryonic cell adhesion. *Nature,* **251**, 522–524.

Balsamo, J. and Lilien, J. (1975), The binding of tissue-specific adhesive molecules to the cell surface. A molecular basis for specificity. *Biochemistry*, **14**, 167–171.

Bell, E.B. and Shand, F.L. (1975), Changes in lymphocyte recirculation and liberation of the adoptive memory response from cellular regulation in the iradiated recipients. *Eur. J. Immunol.,* **5**, 1–7.

Burdick, M.L. (1968), A test of the capacity of chick embryo cells to home after vascular dissemination. *J. exp. Zool.,* **167**, 1–20.

Burdick, M.L. (1970), Cell sorting out according to species in aggregates containing mouse and chick embryonic limb mesoblast cells. *J. exp. Zool.,* **175**, 357–368.

Burdick, M.L. and Steinberg, S. (1969), Embryonic cell adhesiveness: Do species differences exist among warm-blooded vertebrates. *Proc. natn. Acad. Sci. U.S.A.,* **63**, 1169—1173.

Crandall, M.A. and Brock, T.D. (1968), Molecular aspects of specific cell contact. *Science,* **161**, 473—475.

Curtis, A.S.G. (1962), Cell contact and adhesion. *Biol. Rev.,* **37**, 82—129.

Curtis, A.S.G. (1969), The measurement of cell adhesiveness by an absolute method. *J. Embryol. exp. Morph.,* **22**, 305—325.

Curtis, A.S.G. (1970a), Problems and some solutions in the study of cellular aggregation. *Symp. Zool. Soc. Lond.,* **25**, 335—352.

Curtis, A.S.G. (1970b), Re-examination of a supposed case of specific cell adhesion. *Nature,* **226**, 260—261.

Curtis, A.S.G. (1970c), On the occurrence of specific adhesion between cells. *J. Embryol. exp. Morph.,* **23**, 235—272.

Curtis, A.S.G. (1972), Adhesive interactions between organisms. *Symposia Br. Soc. Parasitol,* **10**, 10—21.

Curtis, A.S.G. (1973), Cell adhesion. *Prog. Biophys. Mol. Biol.,* **27**, 315—386.

Curtis, A.S.G. (1974), The specific control of cell positioning. *Arch. Biol.,* **85**, 105—121.

Curtis, A.S.G. (1976), Le positionnement cellulaire et la morphogenese. *Bull. Soc., Zool. France,* **101**, 1—9.

Curtis, A.S.G. (1978), Individuality and graft rejection in sponges *OR* A cellular basis for individuality in sponges. *Biol. and systematics of colonial organisms.* (Rosen, B. ed.), in press. Systematics Assoc.

Curtis, A.S.G. and De Sousa, M.A.B. (1973), Factors influencing adhesion of lymphoid cells. *Nature New Biol.,* **244**, 45—47.

Curtis, A.S.G. and De Sousa, M.A.B. (1975), Lymphocyte interactions and positioning. 1. Adhesive interactions. *Cellul. Immun.,* **19**, 282—297.

Curtis, A.S.G. and Greaves, M.F. (1965), The inhibition of cell aggregation by a pure serum protein. *J. Embryol. exp. Morph.* **13**, 309—326.

Curtis, A.S.G. and Hocking, L.M. (1970), Collision efficiency of equal spherical particles in a shear flow. *Trans. Faraday Soc.* **66**, 1381—1390.

Curtis, A.S.G. and Van De Vyver, G. (1971), The control of cell adhesion in a morphogenetic system. *J. Embryol. exp. Morph.* **26**, 295—312.

De Sousa, M.A.B. and Haston, W.A. (1976), Modulation of B-cell interactions by T cells. *Nature,* **260**, 429—430.

Dubois, R. and Croisille, Y. (1970), Germ-cell line and sexual differentiation in birds. *Phil. Trans. Roy. Soc. Lond. ser. B.,* **259**, 73—89.

Edelstein, B.B. (1970), Cell specific diffusion model of morphogenesis. *J. theor. Biol.,* **30**, 515—532.

Elton, R.A. and Tickle, C.A. (1971), The analysis of spatial distributions in mixed cell populations: a statistical method for detecting sorting out. *J. Embryol. exp. Morph.,* **26**, 135—156.

Ford, W.L. and Simmonds, S.J. (1972), The tempo of lymphocyte recirculation from blood to lymph in the rat. *Cell Tissue Kinet.,* **5**, 175—189.

Ford, W.L., Sedgley, M., Sparshott, S.M. and Smith, M.E. (1976), The migration of lymphocytes across specialized vascular endothelium. II. The contrasting consequences of treating lymphocytes with trypsin and neuraminidase. *Cell Tissue Kinet., 9*, 351–361.

Gerisch, G. (1968), Cell aggregation and differentiation in *Dictyostelium. Current topics in developmental biology, 3*, 157–197.

Goel, N., Campbell, R.D., Gordon, R., Rosen, R., Martinez, H. and Ycas, M. (1970), Self-sorting of isotropic cells. *J. theor. Biol., 28*, 23–68.

Goel, N.S. and Leith, A.G. (1970), Self-sorting of anisotropic cells. *J. theor. Biol., 28*, 469–482.

Greaves, M.F., Owen, J.J.T. and Raff, M.C. (1974), *T and B Lymphocytes: Origins, Properties and Roles in Immune Responses.* Excerpta Medica, Amsterdam.

Harris, A.K. (1976), Is cell sorting caused by differences in the work of intercellular adhesion? A critique of the Steinberg hypothesis. *J. theor. Biol., 61*, 267–285.

Humphreys, T. (1963), Chemical dissolution and *in vitro* reconstruction of sponge cell adhesions. 1. Isolation and functional demonstration of the components involved. *Dev. Biol., 8*, 27–47.

Katz, D.H. and Armerding, D. (1975), Evidence for the control of lymphocyte interactions by gene products of the *I* region of the *H-2* complex. In: *Immune Recognition.* (Rosenthal, A.S., ed.), Academic Press, New York and London, pp. 727–751.

Kondo, K. (1974), Demonstration of a reaggregation inhibitor in sea urchin embryos. *Exp. Res., 86*, 178–181.

Kondo, K. and Sakai, H. (1971), Demonstration and preliminary characterization of reaggregation-promoting substances from embryonic sea urchin cells. *Dev. Growth and Differentiation, 13*, 1–14.

Lilien, J.E. (1968), Specific enhancement of cell aggregation *in vitro Dev. Biol., 17*, 657–678.

Lilien, J.E. (1969), Toward a molecular explanation for specific cell adhesion. *Current topics in Developmental Biol., 4*, 169–193.

Lilien, J.E. and Moscona, A.A. (1967), Cell aggregation: its enhancement by a supernatant from cultures of homologous cells. *Science, 175*, 70–72.

McClay, D.R. (1971), An autoradiographic analysis of the species specificity during sponge cell reaggregation. *Biol. Bull., 141*, 319–330.

McClay, D.R. (1974), Cell aggregation properties of cell surface factors from five species of sponge. *J. exp. Zool., 188*, 89–102.

McClay, D.R. and Hausman, R.E. (1976), Specificity of cell adhesion: differences between normal and hybrid sea urchin cells. *Dev. Biol., 47*, 454–460.

McClay, D.R. and Moscona, A.A. (1974), Purification of the specific cell-aggregating factor from embryonic neural retina cells. *Exp. Cell Res., 87*, 438–443.

McDevitt, H.O. (1975), Genetic control of immunocompetent cell interactions. In: *Immune Recognition,* (Rosenthal, A.S. ed.), Academic Press, New York and London, pp. 621–626.

McGuire, E.J. and Burdick, C.L. (1976), Intercellular adhesive selectivity. 1. An improved assay for the measurement of embryonic chick intercellular adhesion (liver and other tissues). *J. Cell Biol., 68*, 80–89.

Medawar, P.B. (1944), The behaviour and fate of skin autografts and skin homografts in rabbits. *J. Anat.* **78**, 176–199.

Meyer, D.B. (1964), The migration of primordial germ cells in the chick embryo. *Dev. Biol.,* **10**, 154–190.

Moscona, A.A. (1962), Analysis of cell recombinations in experimental synthesis of tissues *in vitro. J. Cell. comp. Physiol.,* Suppl., **60**, 65–80.

Moscona, A.A. (1968), Aggregation of sponge cells: cell-linking macromolecules and their role in the formation of multicellular systems. *In vitro,* **3**, 13–21.

Muller, W.E.G. and Zahn, R.K. (1973), Purification and characterization of a species-specific aggregation factor in sponges. *Exp. Cell Res.,* **80**, 95–104.

Muller, W.E.G., Muller, I., Kurelec, B. and Zahn, R.K. (1976), Species-specific aggregation factor in sponges. IV. Inactivation of the aggregation factor by mucoid cells from another species. *Exp. Cell Res.,* **98**, 32–40.

Muller, W.E.G., Muller, I., Zahn, R.K. and Kurelec, B. (1976), Species-specific aggregation factor in sponges. VI. Aggregation receptor from the cell surface. *J. Cell Sci.,* **21**, 227–241.

Neuwenhuis, C.P. and Ford, W.L. (1976), Comparative migration of B- and T-lymphocytes in the rat spleen and lymph nodes. *Cell. Immunol.,* **23**, 254–267.

Pessac, B. and Defendi, V. (1972), Evidence for distinct aggregation factors and receptors in cells. *Nature New Biol.,* **238**, 13–15.

Pielou, E.C. (1960), A single mechanism to account for regular, random and aggregated populations. *J. Ecol.,* **48**, 575–584.

Raven, C.P. (1967), The distribution of special cytoplasmic differentiations of the egg during early cleavage in *Limnaea stagnalis. Dev. Biol.,* **16**, 407–437.

Roach, S.A. (1968), *The Theory of Random Clumping.* Methuen, London.

Roseman, S. (1970), The synthesis of complex carbohydrates by multiglycosyl-transferase systems and their potential function in intercellular adhesion. *Chem. Phys. Lipids,* **5**, 270–297.

Rosen, S.D., Simpson, D.L., Rose, J.E. and Barondes, S.H. (1974), Carbohydrate-binding protein from *Polysphondylium pallidum* implicated in intercellular adhesion. *Nature* **252**, 128, 149–151.

Roth, S. (1968), Studies on intercellular adhesive selectivity. *Dev. Biol.,* **18**, 602–631.

Roth, S., McGuire, E.J. and Roseman, S. (1971), Evidence for cell-surface glycosyltransferases. Their potential role in cellular recognition *J. Cell Biol.,* **51**, 536–547.

Roth, S.A. and Weston, J.A. (1967), The measurement of intercellular adhesion. *Proc. natn. Acad. Sci. U.S.A.,* **58**, 974–980.

Rutishauser, U., Thiery, J.P., Brackenbury, R., Seal, B.A. and Edelman, G. (1976), Mechanisms of adhesion among cells from neural tissues of the chick embryo. *Proc. natn. Acad. Sci. U.S.A.,* **73**, 577–581.

Sedgley, M. and Ford, W.L. (1976), The migration of lymphocytes across specialized vascular endothelium. 1. The entry of lymphocytes into the isolated mesenteric lymph-node of the rat. *Cell Tissue Kin.,* **9**, 231–243.

Simon, D. (1957), La localisation primaire des cellules germinales dans l'embryon de poulet; preuves experimentales. *C.R. Soc. Biol.* **151**, 1010–1012.

Snell, W.J. (1976a), Mating in *Chlamydomonas*: a system for the study of specific cell adhesion. 1. Ultrastructural and electrophoretic analyses of flagellar surface components involved in adhesion. *J. Cell Biol.*, **68**, 48–69.

Snell, W.J. (1976b), Mating in *Chlamydomonas*: a system for the study of specific cell adhesion. II. A radioactive flagella-binding assay for quantitation of adhesion. *J. Cell Biol.*, **68**, 70–79.

Sprent, J. (1973), Circulating T and B lymphocytes of the mouse. 1. Migratory properties. *Cell. Immunol.*, **7**, 10–39.

Steinberg, M.S. (1962a), On the mechanism of tissue reconstruction by dissociated cells. 1. Population kinetics, differential adhesiveness, and the absence of directed migration. *Proc. natn. Acad. Sci. U.S.A.*, **48**, 1577–1582.

Steinberg, M.S. (1962b), Mechanism of tissue reconstruction by dissociated cells. II. Time-course of events. *Science*, **137**, 762–763.

Steinberg, M.S. (1962c), On the mechanism of tissue reconstruction by dissociated cells. III. Free energy relations and the reorganization of fused, heteronomic tissue fragments. *Proc. natn. Acad. Sci. U.S.A.*, **48**, 1769–1776.

Steinberg, M.S. (1963), Reconstruction of tissues by dissociated cells. *Science*, **141**, 401–408.

Steinberg, M.S. (1970), Does differential adhesion govern self-assembly processes in histogenesis? Equilibrium configurations and the emergence of a hierarchy among populations of embryonic cells. *J. exp. Zool.*, **173**, 395–434.

Steinberg, M.S. (1975), Adhesion-guided multicellular assembly: a commentary upon the postulates, real and imagined, of the differential adhesion hypothesis, with special attention to computer simulations of cell sorting. *J. theor. Biol.*, **55**, 431–443.

Swift, D.L. and Friedlander, S.K. (1964), The coagulation of hydrosols by Brownian motion and laminar shear flow. *J. Colloid Sci.*, **19**, 621–647.

Takahashi, K. and Okada, T.S. (1971), Separation of two factors affecting the aggregation kinetics from the conditioned medium. *Dev. Growth and Differentiation*, **13**, 15–24.

Townes, P.L. and Holtfreter, J. (1955), Directed movements and selective adhesion of embryonic amphibian cells. *J. exp. Zool.*, **128**, 53–120.

Turner, R.S. and Burger, M.M. (1973), Involvement of a carbohydrate group in the active site for surface guided reassociation of animal cells. *Nature*, **244**, 509–510.

Tuler, A. (1946), An auto-antibody concept of cell structure, growth and differentiation. *Growth*, **10**, 7–19.

Van De Vyver, G. (1970), La non confluence intraspecifique chez les spongiaires et la notion d'individu. *Annales Embryol. Morphog.*, **3**, 251–262.

Van De Vyver, G. (1975), Phenomena of cellular recognition in sponges. *Current Topics in Dev. Biol.*, **10**, 123–140.

Walther, B.T., Ohman, R. and Roseman, S. (1973), A quantitative assay for intercellular adhesion. *Proc. natn. Acad. Sci. U.S.A.*, **70**, 1569–1573.

Walther, B. Rausch, B. and Roseman, S. (1976), Sequential reactions in intercellular adhesion. *J. Cell Biol.*, **70**, 70a.

Weinbaum, G. and Burger, M.M. (1973), Two component system for surface-guided reassociation of animal cells. *Nature*, **244**, 510–512.

Weiss, P. (1947), The problem of specificity in growth and development. *Yale J. Biol. and Med.*, **19**, 235–278.

Willis, R.A. (1952), *The Spread of Tumours in the Human Body*. Butterworth, London.

Wolpert, L. (1969), Positional information and the spatial pattern of cellular differentiation. *J. theor. Biol.*, **25**, 1–47.

Wood, S. (1964), Experimental studies of the intravascular dissemination of ascitic V2 carcinoma cells in the rabbit, with special reference to fibrinogen and fibrinolytic agents. *Bull. Swiss. Acad. Med. Sci.*, **20**, 92–121.

Wourms, J.P. (1972), The developmental biology of annual fishes. II. Naturally occurring dispersion and reaggregation of blastomeres during the development of annual fish eggs. *J. exp. Zool.*, **182**, 169–200.

Zeidman, I. (1961), The fate of circulating tumor cells. I. Passage of cells through capillaries. *Cancer Res.*, **21**, 38–39.

Part 3: Cell Adhesion in 'Model Systems'

6 Sponge Cell Adhesions

R. S. TURNER Jr.

Acknowledgements

I would like to express my gratitude to Dr Max M. Burger for the opportunity to work on sponge cell adhesion and for his enthusiastic guidance throughout my association with him. Additional thanks are due to Dr Tom Humphreys, Dr Tracy Simpson and Dr George Weinbaum for their comments on one draft of this chapter.

The author was a Fellow of the Damon Runyon Foundation for Cancer Research and received additional support from a grant from the Swiss National Fund while he participated in the research cited above. A grant from the Biology Department at Wesleyan University supported the preparation of this manuscript.

Specificity of Embryological Interactions
(*Receptors and Recognition,* Series B, Volume 4)
Edited by D.R. Garrod
Published in 1978 by Chapman and Hall, 11 New Fetter Lane, London EC4P 4EE
© Chapman and Hall

6.1 INTRODUCTION

Seventy years ago, H.V. Wilson demonstrated that marine sponges could be dissociated into single cell suspensions simply by mincing the sponge and squeezing the pieces of sponge through bolting silk (Wilson, 1910). The material that passed through the cloth consisted of a suspension of predominantly single cells. When cells from an orange marine sponge (*Microciona prolifera*) were placed in culture on coverslips or slides, the cells attached to the glass and eventually reconstituted a small, functional sponge (Wilson, 1907). When *Microciona prolifera* cells were mixed with cells from a yellow marine sponge, *Cliona celata,* two sets of aggregates, one orange and one yellow, were formed (Wilson, 1907). These results indicate the existence of a species-specific recognition system in sponge cells.

During the last seventy years a variety of studies have been conducted to determine the cellular and biochemical nature of this recognition system. This review will summarize the majority of those studies concerned with analyses of cell recognition in intact sponges and sponge tissue grafts, observations of stationary cultures of dissociated cells, studies of rotation-mediated reaggregation of single cells and biochemical analyses of aggregation factors. In the concluding section, some comments will be offered on the possible relevance of these studies to the biology of the intact sponge.

6.2 CELL RECOGNITION IN INTACT SPONGES

Naturally occurring self-recognition might be expected to operate between free-swimming sponge larvae, between hatching gemmules and between outgrowths from established sponges. All of these cases have been examined. Two larvae will not adhere unless they are products of the same individual sponge. After two larvae do adhere, they can merge and form a single new sponge (Levi, 1956; Van de Vyver, 1970).

The fusion of two gemmules to form one sponge has been studied extensively in *Ephydatia fluviatilis* by Van de Vyver (1970, 1975). She found that gemmules from some pairs of individual sponges can fuse upon hatching while gemmules from other pairs cannot. Van de Vyver used the inability of hatching gemmules to fuse as the basis for distinguishing several strains of *E. fluviatilis.* Therefore, the fusion of hatching gemmules is strain-specific in *Ephydatia fluviatilis.*

The outgrowth formed from sponge explants has been described in several species (Wilson and Penney, 1930; Simpson, 1963, 1973; Harrison, 1974). Simpson (1973) reported that two explants from the same *Microciona prolifera* individual formed outgrowths that merged, while outgrowths from two different individuals did not. A museum specimen has been reported to contain interspersed gemmules and spicules from *Anheteromeyenia ryderi* and *Enapicus fragilis,* which the author contends establishes an intergeneric sponge mixture. (Smith, 1976). However, in view of the ability of cells from some species of sponge to lyse cells from other species (Galtsoff, 1925b, 1929), it is possible that this specimen arose by the invasion and lysis of one species by the other.

The weight of these observations of cell recognition in intact sponges suggests that it is possible for sponge cell recognition to occur at the level of the individual sponge. We currently lack enough information to predict whether recognition and subsequent merger will occur in a specific case except for the strain specificity of *Ephydatia fluviatilis* gemmules. The evidence summarized in the next four sections will indicate that the cellular and biochemical basis for species-specific sponge cell recognition is under intensive study. The results just described emphasize that the cellular and biochemical studies of species-specific cell associations will eventually have to be integrated with specific cell recognition at both the strain and the individual level in intact sponges.

6.3 CELL RECOGNITION IN TISSUE GRAFTS

Some studies of dissociated sponge cells have included as controls an examination of the species specificity of graft rejection. Both Moscona (1968) and McClay (1974) have reported that homografts in which donor and recipient were of the same species were accepted in all cases, while heterografts were rejected.

Moscona found that all homospecific grafts between different *Microciona prolifera* individuals from different locations and of different ages were accepted within 12 hours. At this time, the cells from donor and host were intermingled. Such grafts survived up to three weeks.

Heterospecific grafts between *M. prolifera* and *Haliclona occulata* were rejected with only the skeleton of the graft remaining. McClay's heterografts either displayed necrosis at the interface between donor and host or the heterografts fell away from the host when the pins that held donor and host together were removed (McClay, 1974). Paris, in contrast to both Moscona and McClay, observed heterograft fusion accompanied by intermingling of cells and extracellular components in a study of *Tethya lyncurium* and *Suberites domuncula* (Paris, 1960).

The results with *Microciona prolifera* homografts differ from the observations of *M. prolifera* explant outgrowths since all homografts were accepted while outgrowths from some pairs of explants from different sponges did not fuse (Simpson, 1973). In fact, what we interpret as a zone of necrosis, perhaps similar to that in some of McClays's heterografts, is visible between *Microciona prolifera* outgrowths that do not merge (Fig. 14, Simpson, 1973). This analysis suggests that the rejection mechanism may be similar in both outgrowths and tissue grafts, since a zone of necrosis occurs in both cases. However, the recognition mechanism may differ in

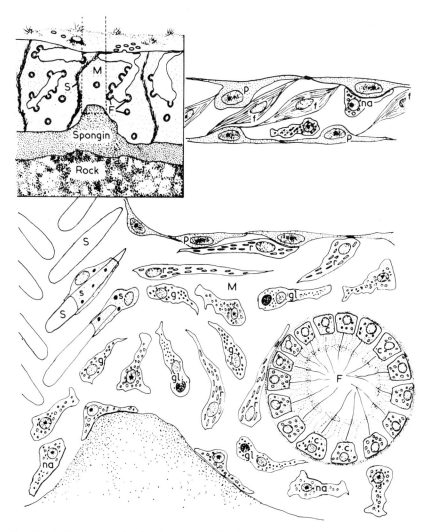

Fig. 6.1 Cellular organization in a generalized sponge. The area outlined by the dotted line in the inset approximates that shown in the expanded cross-section. The cells and their organization are based on the observations in Simpson (1963, 1968) and Smith and Lauritis (1969). The diversity of the cell types present in the mesohyl and the contrast between the loosely organized mesohyl and the structured organization of both the pinacoderm and the flagellated chambers should be noted.

c = choanocyte, f = fiber cell, F = flagellated chamber, g = grey cell, gl = globoferous cell, M = mesohyl, na = nucleolate ameboid archeocytes, p = pinacocyte, r = rhabdiferous cell, s = scelerocyte, S = spicule.

The author would like to express his appreciation to Dr Spencer J. Berry of the Wesleyan University Biology Department for drawing the figure.

outgrowths and in grafts, since the grafts between individual *M. prolifera* were never rejected while rejection between individuals did occur in the outgrowth studies. Simpson (1963) has pointed out that in outgrowths the pinacocytes come into apposition, while the mesohyl is the probable site of contact between donor and host in grafts. The implication of these considerations is that a pinacocyte-mediated recognition process may possess greater specificity than a recognition process mediated by the mesohyl, at least in *Microciona prolifera,* (see Fig. 6.1 for a diagram of the organization of a generalized sponge).

6.4　CELL RECOGNITION IN STATIONARY CULTURES

The experiments discussed in this section resemble Wilson's original study since they were all conducted by compressing sponges through bolting cloth to obtain cells in suspension. The cells were then allowed to settle onto a substratum and their behavior observed.

The sponge cell suspension contain several types of cells. Amoeboid, nucleolate archeocytes and, in some species, grey cells are present in the loosely organized mesohyl of the intact sponge and these cell types are quite numerous in the cell suspensions (Wilson and Penney, 1930; Galtsoff, 1923, 1925a, de Laubenfels, 1934; Borojevic and Levi, 1964; Mookerjee and Garguly, 1964). In contrast, the pinacocytes from the highly organized epithelial pinacoderm of the intact sponge make a much more variable contribution to the cell types present in the suspensions, ranging from a very small (Wilson and Penney, 1930; Borojevic and Levi, 1964; Mookerjee and Garguly, 1964) to a large (Galtsoff, 1923, 1925a; de Laubenfels, 1934) proportion of the cells present. This variability is not unexpected, since the mechanical force necessary to disrupt the pinacoderm may vary among different species of sponge. Clusters of choanocytes or collar cells form flagellated chambers in the intact sponge which, like the pinacoderm, is an organized epithelium. However, unlike the variable proportion of pinacocytes in the cell suspensions, choanocytes are quite numerous (Wilson and Penney, 1930; Galtsoff, 1923, 1926; de Laubenfels, 1934; Borojevic and Levi, 1964; Mookerjee and Ganguly, 1964) probably because intact flagellated chambers may pass through the bolting cloth.

The most significant feature of the cell suspensions is that their composition from a given species may change. Agrell (1951) has shown that subjecting intact sponges to starvation causes the cell suspensions to contain fewer archeocytes and more choanocytes. Wilson and Penney (1930) and de Laubenfels (1927) both report seasonal variations in the proportion of embryo cells in their cell suspensions. Simpson's (1968) careful recording of the seasonal appearance of eggs, sperm, and embryos in intact *Microciona prolifera* indicates that the presence of embryos in the cell suspensions reflects changes occurring in the sponge. Since archeocytes and/or choanocytes are the probable source of eggs, sperm and nurse cells and since, as discussed below, archeocytes are essential for cell reassociation, it should not be surprising that seasonal and age-related changes in cell reassociations occur (Van de Vyver, 1975; deLaubenfels, 1927). Furthermore, although this question has not been examined in detail, it is possible that fluctuations in the cell types present in the cell suspensions may also affect the results of the studies of rotation-mediated reaggregation and aggregation factor activity that are discussed in the next two sections.

Once the cell suspension has been prepared, the stationary cultures are observed after allowing the cells to settle onto the substratum. It is generally agreed that cell contacts occur in large part as a result of the mobility of the archeocytes (Galtsoff, 1923; 1925a; 1929; Mookerjee and Ganguly, 1964; Ganguly, 1960; Sindelar and Burnett, 1967). In monospecific cultures the archeocytes are very motile and collect the cells they contact as they migrate. Galtsoff studied bispecific suspensions of cells from *Microciona prolifera* and *Cliona celata* (1925a) and observed that archeocyte motility was inhibited by heterospecific cells. He later made similar observations with several Bermudan sponges (Galtsoff, 1929). Galtsoff concluded that mutual inhibition of archeocyte movement partially accounts for the formation of mono-specific aggregates.

In several particular bispecific mixtures, Galtsoff, (1923, 1925a, 1929) found that motile archeocytes did not adhere to and collect heterospecific cells. This basic observation has been extended by studies of bispecific cell suspensions from *Microciona prolifera* and *Haliclona occulata* using time lapse cinematography (Sindelar and Burnett, 1967) and determinations of electrical coupling (Lowenstein, 1967). The time-lapse study indicated that heterospecific cell contacts did not produce effective adhesions (Sindelar and Burnett, 1967) and the electrophysiological measurements determined that heterospecific contacts did not produce ionic communication between the cells (Lowenstein, 1967). In contrast, homospecific contacts led to effective adhesions (Sindelar and Burnett, 1967) and ionic communication (Lowenstein, 1967) the latter occurring within 10 minutes of the initial contact.

The conclusion from these observations is that motile archeocytes collect homospecific cells preferentially, thereby producing monospecific aggregates which may then reconstitute intact sponges. This conclusion implies that initial cell adhesions accompany or are directed by a cell recognition event.

The experiments described in the rest of this section have been used in support of the opposite sequence of events, i.e., that cell adhesion formation precedes the cell recognition event. Before describing these studies, we would like to indicate that the results of some of these studies could be produced either by non-specific initial adhesions followed by specific cell sorting or by specific initial adhesions followed by non-specific aggregate fusion (Fig. 6.2).

The most straightforward demonstration that adhesion precedes recognition occurs when initial bispecific aggregates form and the cells subsequently either sort out in a species-specific manner within the aggregate (Fig. 6.2, 1 to 3a) or the species-specific sorting occurs to such an extent that monospecific aggregates are eventually

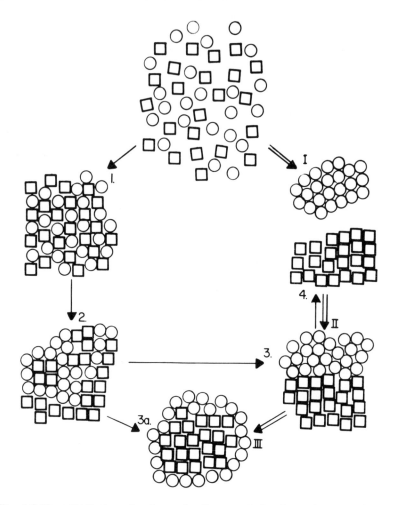

Fig. 6.2 Hypothetical mechanisms of cell reassociation in stationary cultures. (Circles and squares represent cells from two different species of sponge).

The sequences 1 to 3a and 1 to 4 indicate the events possible if cell recognition occurs after cell adhesion. The sequence I to III may be anticipated if cell recognition accompanies or directs the adhesive events. Additional discussion is provided in the text.

formed (Fig. 6.2, 1 to 4). In the case of cell sorting within the aggregates, two discrete areas exist, such that, in the extreme case, the cells of one species form a complete 'coating' over a central 'core' of cells of the second species (Fig. 6.2, 3a). However, Steinberg (1970) has shown that this configuration can also result from the apposition of two monospecific aggregates that contain different cell types.

Although a similar experiment has not been performed with sponge aggregates, it is hypothetically possible for two homospecific aggregates of different species to adhere and form a single heterospecific aggregate (Fig. 6.2, I-III). If this did occur, homospecific recognition would have preceded or directed cell adhesion. This analysis indicates that it is possible to generate both monospecific and bispecific aggregates either by specific initial adhesions followed by aggregate fusion or by non-specific initial adhesions followed by cell sorting. Therefore, it is necessary to monitor the early stages of aggregation to determine whether the initial adhesions are specific or non-specific and also to examine the distribution of the cells within bispecific aggregates to determine if cell sorting has occurred. The same determinations are obviously required when monospecific aggregates are formed in order to decide whether they were the result of specific adhesions or of non-specific adhesions followed by extensive cell sorting. Most of the studies we will now describe have assumed that bispecific aggregates can arise only from non-specific initial adhesions and that monospecific aggregates arise only from specific initial adhesions.

A case in point is the report by Sara *et al.* (1966) that some bispecific cell mixtures result in the formation of predominantly bispecific aggregates, while other combinations result in both monospecific and bispecific aggregates and still other mixtures form primarily monospecific aggregates. Since aggregate formation was observed only after 20 hours of incubation, it is not possible to decide whether the initial cell adhesions were solely homospecific or whether heterospecific adhesions were also formed. Both possibilities exist, even for Sara's bispecific aggregates, since the cells within these aggregates occur in two separate, apparently homogeneous, monospecific populations. As we have just indicated, this configuration could arise either by the fusion of small monospecific aggregates or by the formation of heterospecific adhesions between individual cells followed by cell sorting within the aggregate.

Bispecific aggregates have been produced experimentally by deLaubenfels (deLaubenfels, 1927, 1928, 1934). Using different species than the investigators cited so far, he observed the formation of monospecific aggregates when the cell suspensions were prepared separately and the monospecific cell suspensions were then mixed. Small monospecific aggregates that were forced together with glass needles underwent species-specific sorting, thus providing further evidence for the preferential formation of homospecific adhesions. However, when minced sponge tissue from two species was mixed before being forced though the bolting cloth, the resultant bispecific cell suspension reportedly formed bispecific aggregates. deLaubenfels claims that this procedure reconstitutes bispecific conglomerates that, rarely, form functional flagellated chambers but that bispecific aggregates do not survive as well as the corresponding monospecific aggregates. Since flagellated chambers occured only where one species' cells were in excess, these bispecific aggregates probably arise only after extensive species-specific cell sorting.

In all of the above experiments sponges were mechanically dissociated by being forced through bolting cloth. In experiments discussed in detail in Section 6.6,

Humphreys (1963) dissociated sponges with calcium-free sea water washes and obtained a soluble molecule that is required for homospecific adhesion formation. Curtis (1962) and Johns *et al.* (1971) dissociated sponges with EDTA washes, rather than calcium-free sea water washes, and observed the behavior of the cells in stationary cultures.

Curtis formed bispecific mixtures from four different species (Curtis, 1962). When the cells were mixed at the start of the reassociation period, monospecific aggregates were formed in three cases, while in the final case, completely bispecific aggregates resulted. Curtis was able to manipulate the extent of bispecific aggregation that occurred by pre-incubating one cell suspension before adding the second. Based on this observation, he concluded that the formation of homospecific or heterospecific aggregates depended on differences in the time required for cells from different species of sponge to recover from the dissociation procedure and regain the ability to form homospecific adhesions. However, since the cultures were observed only after 20 hours of reassociation, it is not possible to decide whether the initial adhesions were heterospecific or homospecific.

Like Curtis (1962), Johns *et al.* (1971) used EDTA washes during dissociation and monitored aggregate specificity only after almost 20 hours of reassociation. In addition, they isolated archeocytes and mucoid cells from cell suspensions obtained from two different species and observed reassociation of all possible combinations of cells from both species. The mucoid cells stained with acidic Alcian Blue, indicating that they contain acid mucopolysaccharide and thus may be identical to grey cells, rhabdiferous cells or globoferous cells. (Simpson, 1963). Since the mucoid cells are found near choanocytes in Ficoll gradients (Johns *et al.*, 1971, we feel that the mucoid cells are too small to be either rhabdiferous or globoferous cells and will therefore equate the mucoid cells with grey cells.

Johns *et al.* (1971) found that isolated grey (mucoid) cells do not reaggregate, while isolated archeocytes do reaggregate. Furthermore, when various combinations of grey (mucoid) cells and archeocytes from the two species were examined, it became obvious that the archeocytes were mandatory for reaggregation, but that monospecific aggregate formation required grey (mucoid) cells. This experiment has not been repeated, but it suggests that adhesive contacts and *homospecific* adhesive contacts are due to two different types of cells.

The use of EDTA washes in the dissociation procedure will be the subject of further comment in subsequent sections. The results of both Curtis (1962) and Johns *et al.* (1971) must be interpreted after considering both the adverse effect of EDTA on soluble aggregation factors (Section 6.6) and the increased leakiness that sponge cells exhibit in low concentrations of calcium (Lowenstein, 1967). The use of EDTA as a dissociating agent by Curtis (1962) and by Johns *et al.* (1971) suggests that their results and conclusions may not be directly comparable to the results obtained with mechanically dissociated cells that were presented earlier in this section.

These studies of sponge cells in stationary culture indicate that motile archeocytes and the acid mucopolysaccharide containing grey (mucoid) cells are necessary for species-specific aggregate formation. We have pointed out that species-specific aggregates may be produced either by the formation of species initial cell contacts or by the formation of non-specific initial contacts followed by cell sorting within the aggregates on a species-specific basis. The former mechanism, which suggests that cell recognition is closely associated with the formation of cell adhesion, appears to operate in the bispecific mixture of *Microciona prolifera* and *Haliclona occulata.* The other bispecific mixtures have not been clearly demonstrated to produce species-specific initial cell adhesions. Species-specific cell recognition occurs after cell adhesion in some of the mixtures and generates species-specific aggregates in those mixtures.

6.5 CELL RECOGNITION IN ROTATION-MEDIATED REAGGREGATION

A major advance in studies of cell associations occurred with the introduction of rotation-mediated reaggregation (Gerisch, 1960; Moscona, 1961) which involves placing the cell suspension in an assay vessel that is subjected to agitation. This substitutes cell contacts produced by random collisions due to agitation for contacts produced by cell motility. Eliminating the dependence on cell movement should simplify the cellular behavior necessary for reassociation and consequently simplify analyses of cell adhesion.

Humphreys (1963) used rotation-mediated reaggregation to demonstrate the existence of species-specific aggregation-promoting factors (see Section 6.6). He found that dissociation in calcium-free sea water rather than artificial sea water decreased *Microciona prolifera* cell adhesiveness. Calcium-free sea water-dissociated (CF) cells reaggregated more slowly at room temperature than mechanically dissociated (MD) cells and MD cells reaggregated at low temperature, while CF cells did not. These differences between MD and CF cells did not arise from damage to the CF cells, since the CF cells were still capable of forming functional sponges in stationary cultures. Furthermore, when CF cells at low temperature were shifted to room temperature, they reaggregated as if they have been at room temperature continuously (Humphreys, 1963). Reaggregation of CF cells depended upon the presence of Ca^{2+} and Mg^{2+} or Ca^{2+} and Sr^{2+} in the reassociation medium.

Histological examination of aggregates revealed that reaggregation occurred in two steps: temperature-independent cell adhesion and temperature-dependent aggregate compaction. The temperature independence of adhesion formation probably reflects the absence of a requirement for metabolic energy, since protein synthesis is not required for adhesion (Humphreys, 1965) and since formalin-fixed MD cells will form adhesions (Moscona, 1963). Cell adhesion is a relatively rapid process that is completed in less than 3 hours. Aggregate compaction or 'firming up', which is completed in about 12 hours, involves cell movement within the aggregate and flattening of the cells on the aggregate surface. MD cell adhesion, the initial step in aggregate formation, was species-specific in two bispecific mixtures; those between *Microciona prolifera* and *Haliclona occulata* and between *Haliclona occulata* and *Halichondria panicea.*

Since not all sponge species can be dissociated into cells capable of rotation-mediated reaggregation (Humphreys, 1969) and since very few sponge species reconstitute functional sponges from dissociated cells, it appears that different sponge species vary in their response to dissociation. Therefore, it is not surprising that species-specific cell adhesion is not a property of all bispecific mixtures (Humphreys, 1969). Humphreys screened a large number of bispecific combinations, and concluded that, in most combinations initial cell adhesions are non-specific. He concludes that subsequent species-specific cell sorting is responsible for almost all cases of what is commonly referred to as species-specific reaggregation. He found that only the bispecific mixtures *Haliclona occulata* – *Microciona prolifera* and *Haliclona occulata* – *Halichondria panicea* formed species-specific *initial* cell adhesion.

Support for species-specific initial cell adhesions in the *Microciona prolifera* – *Haliclona occulata* combination is provided by the time-lapse cinematographic (Sindelar and Burnett, 1967) and the electrical coupling (Lowenstein, 1967) experiments described in the previous section. Moscona (1963) has also shown that cells from these two species reaggregate completely specifically. Curtis has observed bispecific aggregate formation in mixtures of *Haliclona occulata* and *Halichondria panicea* cells during the first 40 minutes of reaggregation. Though his observation apparently contradicts Humphreys' 1963 results, as we previously noted, Curtis' cells were obtained using EDTA washes (Curtis, 1962, 1970a). EDTA causes irreversible dissociation of sponge aggregation-promoting materials (Moscona, 1963; Cauldwell *et al.,* 1973) and also extracts some cell membrane components (Bhakdi *et al.,* 1974). Humphreys has pointed out (1970) that EDTA-dissociated cells do, in fact, reaggregate more slowly than cells dissociated in calcium-free sea water. This evidence implies that EDTA dissociation produces cell suspensions that are not identical to those produced by calcium-free sea water, thereby suggesting that the lack of agreement between Curtis' and Humphreys' results may exist only because the cell suspensions were prepared differently. The available evidence supports Humphreys' observation that the initial cell adhesions formed between *Microciona prolifera* and *Haliclona occulata* or between *Haliclona occulata* and *Halichondria panicea* are species-specific. Therefore, in these two cases cell recognition preceeds and/or directs the formation of cell adhesions.

McClay (1971) has studied rotation-mediated sponge cell associations in a large number of Bermudan species. His experimental procedures differ in two respects from those used in the studies described so far: he employed radioactive labeling to distinguish cells of different species and the collecting aggregate assay to measure heterospecific and homospecific cell associations.

The previously cited studies distinguished between cells of different species either by the natural color of the sponge (Wilson, 1907, 1910) or by staining one species with a vital dye (Sara *et al.,* 1966). Both of these criteria may be misleading, since, as Curtis (1962) has pointed out, they are only totally accurate if all cells in the sponge are pigmented or if no loss of stain occurs. In fact, in most species only specific cell types are pigmented (Wilson and Penney, 1930). The use of vital dyes is probably more legitimate, though studies of the loss of stain have not been reported. McClay (1971) has compared the distribution of cells labeled with radioactive thymidine to the predominant color of individual aggregates. The distribution of labeled cells present when labeled cells of one species are mixed with unlabeled cells of a second species indicated that initial non-specific aggregation occurs. In contrast, the color of each aggregate after one to two hours was that of either one species or the other, suggesting that specific adhesion occurred. This comparison of aggregate color with the radioactive label distribution suggests that the two marking methods may lead to two different conclusions with respect to the specificity of the initial adhesions.

The collecting aggregate assay that McClay used was developed by Roth and Weston (1967) to measure the specificity of cell adhesions. In this assay the number of radioactive cells in a single cell suspension that bind to preformed aggregates of unlabeled cells from the same tissue measures homotypic adhesions, while the number of labeled cells bound to aggregates from a different tissue measures heterotypic adhesiveness. The ratio of homotypic to heterotypic binding is then used as an estimate of the specificity of cell adhesions.

McClay (1971) used autoradiography in conjunction with the collecting aggregate assay to analyze the species-specificity of sponge cell adhesions. Only in homospecific combinations of cells and aggregates were radioisotope-labeled cells from the suspension found in the interior of the unlabeled aggregate. In one hetersospecific case, that of *Haliclona viridis* and *Tedania ignis,* no radioactive cells bound to the aggregates. Six other bispecific combinations usually resulted in a layer of radioactive cells at the surface of the aggregate and no radioactive cells in the aggregate interior.

McClay interprets all of his results as indicating that non-specific adhesions were formed and the cells then sorted out. We would agree in those six bispecific combinations in which heterospecific binding occurs. In those cases, adhesion is non-specific. The absence of heterospecific cells in the aggregate interior shows that heterospecific mixing cannot occur, conceivably because those heterospecific cells that attach to the aggregate 'sort out' homospecifically every time they initiate an attempted movement to the aggregate interior. We differ with McClay's interpretation of the *Haliclona viridis* and *Tedania ignis* combination, since we feel that the absence of radioactive cells on the aggregate surface indicates that heterospecific adhesions do not occur in this mixture.

MacLennan (1969a, 1974; MacLennan and Dodd, 1976) has conducted immunochemical studies of sponge cell surfaces and aggregation factors that will be discussed further in the next section. These studies include several analyses of the species-specificity of rotation-mediated reaggregation. Like Humphreys (1969), he has found that there are a number of species that are capable of reaggregating. When he examined 16 hour old aggregates that do form in bispecific cell suspensions, he found that no bispecific mixing had occurred and concluded that reaggregation is species-specific. However, as we pointed out above, it is necessary to determine the specificity of cell adhesion during the early phases of reaggregation to be certain that MacLennan's results are not due to non-specific adhesion followed by cell sorting.

In addition to the bispecific combination *Haliclona occulata* and *Halichondria panicea* that we mentioned earlier (Curtis, 1970a), Curtis has also examined the bispecific combinations formed between *Microciona fallax, Halichondria panicea* and *Suberites ficus* (1970b). All possible combinations of the last three species formed non-specific adhesions during the first forty minutes of reaggregation. However, the cells were obtained by EDTA dissociation, which, as we mentioned above, may affect the membranes, aggregation factors and adhesive specificity of the cells.

Curtis has also collaborated with Van de Vyver in a study of the strain specificity of cell associations in *Ephydatia fluviatilis*. As indicated in Section 6.2, Van de Vyver classifies individual *E. fluviatilis* into different strains on the basis of the inability of two gemmules from different strains to fuse and produce a single sponge. When cells from EDTA-dissociated gemmules of different strains were mixed, the initial aggregation was non-specific (Curtis and Van de Vyver, 1971) and was followed by cell sorting on a strain-specific basis. This result suggests that the lack of gemmule fusion between strains is not due to strain-specific cell adhesion. However, it is possible that the cell suspension whose adhesiveness was assayed do not represent the cells that interact during gemmule fusion. Apparently, gemmule fusion is mediated by the pinacoderm of the gemmules (Van de Vyver, 1975; Curtis and Van de Vyver, 1971). Since other investigators report that the proportion of pinacocytes in sponge cell suspensions can be quite variable (Section 6.4) and since Curtis and Van de Vyver do not analyze the cell types present in their suspensions, it is still possible that strain-specific gemmule fusion involves strain-specific pinacocyte adhesions.

Investigations of *Geodia cydonium* aggregation have determined that this species aggregates in two steps; cell adhesion to form small primary aggregates and primary aggregate fusion to form large secondary aggregates (Müller and Zahn, 1973). When *Ircinia muscarum* cells were mixed with *Geodia cydonium* cells, the secondary phase of *Geodia cydonium* aggregation was inhibited and almost all of the inhibitory activity was associated with *Ircinia muscarum* mucoid cells (Müller *et al.,* 1976a). This observation agrees quite well with the results of Johns *et al.* 1971) that grey (mucoid) cells are required for homospecific cell associations. It is possible that, if the study by Johns *et al.* (1971) of *Ophlitaspongia seriata* and *Halichondria panicea* cells in stationary culture was repeated using rotation mediated reaggregation, either the grey cells from *Ophilitaspongia seriata* would inhibit *Halichondria panicea* reaggregation or the reverse would be true. These studies of Müller and Zahn (1973, Müller *et al.,* 1976a) and of Johns *et al.* (1971) emphasize the unequal contribution of different cell types to sponge cell reaggregation.

The use of rotation-mediated reaggregation to examine sponge cell adhesions has produced results that do not indicate any one mechanism of cell adhesion that is common to all cases examined. Since this is so, it is also impossible to describe a universal mechanism of generating the specificity of sponge cell interactions that has been observed in some cases. However, in spite of several significant differences in experimental design and in spite of the fact that only one case exists in which two investigators have studied the same bispecific mixture, a few very broad generalizations are possible.

(1) Not all species can reaggregate and very few of those which reaggregate are reported to then form functional sponges.
(2) Most bispecific combinations aggregate by forming heterospecific cell adhesions and then undergoing varying degrees of species-specific cell sorting.
(3) Not all cell types contribute equally to aggregate formation.
(4) CF cells differ from MD cells in the initial phase of reaggregation and EDTA cells do not express the same reaggregation specificities as either CF or MD cells.
(5) Only three bispecific mixtures exist in which species-specific recognition accompanies cell adhesion: *Microciona prolifera* and *Haliclona occulata* (Humphreys, 1963), *Haliclona occulata* and *Halichondria panicea* (Humphreys, 1963), and *Haliclona viridis* and *Tedania ignis* (McClay, 1971).

6.6 BIOCHEMISTRY OF CELL RECOGNITION: AGGREGATION FACTORS

In the previous section, we discussed Humphreys' observation that the reaggregation of CF cells* from *Microciona prolifera, Haliclona occulata* and *Halichondria panicea* is qualitatively different from the reaggregation of MD cells from the same species. These observations indicated that CF cells had been reversibly depleted of cell adhesion components that remained on MD cells (Humphreys, 1963, 1969; Moscona, 1963). CF dissociation washes from *Microciona prolifera* and *Haliclona occulata,* but not from *Halichondria panicea,* restored the reaggregation capabilities of MD cells to the homospecific CF cells. The inability of *Halichondria panicea* to restore MD cell reaggregation to CF cells causes the *Haliclona occulata – Microciona prolifera* combination to become absolutely unique. This is the only bispecific combination in which it has been demonstrated that cell recognition accompanies cell adhesion and in which both species' CF dissociation washes restore MD cell reaggregation to CF cells. These two facts mean that it is possible to obtain the molecule (or molecules) responsible for specific adhesion and to assay both the adhesion promoting activity and the species-specificity of the molecule(s). Similar biochemical analyses have been conducted using other bispecific combinations

---

* Abbreviations: AF = aggregation factor; BP = baseplate; CF = cells obtained by dissociating the sponge in calcium-free sea water; HY = CF cells that have been washed with a dilute NaCl solution to remove their baseplate; MD cells = cells obtained by dissociating the sponge in complete sea water.

(see below) but either the results or their interpretations, or both, differ from the straightforward analysis possible with the *Microciona prolifera — Haliclona occulata* mixture because of the complications inherent in other bispecific mixtures.

It should be emphasized that, though the active component of the CF dissociation wash has been named aggregation factor (AF), its activity is not confined to enhancing the rate or extent of aggregate formation. Since the aggregation capabilities of MD cells have been determined and since AF restores these characteristics to CF cells, the adhesive properties of the cells have been defined both before (MD) and after (CF) they have been stripped of their adhesive component (AF). Only this determination of the adhesiveness of *both* MD and CF cells insures that biochemical studies of AF and its interaction with CF cells are analyzing the mechanism of cell adhesion.

The biological properties of *Microciona prolifera* aggregation factor (AF) are consistent with its proposed role in species-specific cell adhesion. *Microciona* AF preferentially enhances the reaggregation of *Microciona prolifera* CF cells at $5°C$ when tested either with separate suspensions of *Microciona prolifera* CF cells and *Haliclona occulata* CF cells or with a mixture of CF cells from these two species (Humphreys, 1963), which clearly demonstrates its species-specific action. This effect at $5°C$ is on cell adhesion and not aggregate 'compacting' since neither MD cells nor CF cells with added AF undergo compacting at $5°C$ (Humphreys, 1963). Formalin-fixed CF cells are not adhesive unless AF is added to the cells (Moscona, 1963) and, like low temperature, fixation prevents aggregate compaction (Moscona, 1968). AF-mediated adhesion of either fixed or unfixed CF cells is associated with the species-specific loss of AF activity from the medium, suggesting that the cells bind AF rather than becoming adhesive due to some transitory action of or association with AF (Moscona, 1968). AF does not appear to be a cell lysis by-product, since it is not released by homogenization (Humphreys, 1963). Finally, AF release and cell dissociation both occur in the absence of calcium and no detectable AF is released during the preparation of MD cells or by washing MD cells with complete sea water (Humphreys, 1963). However, MD cells can react with AF and both MD and CF cells reaggregate more rapidly and form larger 24 hour old aggregates in the presence, than in the absence, of AF (Moscona, 1968). This last observation may indicate either that MD cells have lost some AF and can therefore bind additional AF from the medium or that the AF on the MD cells self-associates with AF in the medium.

Two groups have purified *Microciona* AF and published electron micrographs of their preparations (Margoliash *et al.*, 1965; Henkart *et al.*, 1973; Humphreys *et al.*, 1975). In one case, crude aggregation factor preparations contained 20 to 25 Å diameter particles which were sometimes seen to be associated in aggregates 200 to 2000 Å in diameter. The identification of these particles or aggregates with the AF activity is uncertain because no attempt was reported to correlate experimentally induced changes in AF activity with changes in the ultrastructure of the particles or aggregates (Margoliash *et al.*, 1965). *Microciona parthena* AF preparations of somewhat greater purity contained structures resembling a sunburst; i.e., a central circular fiber with straight rods beginning at the circle and extending radially outward from it. The fiber that constitutes the central circle and the arms is about 45 Å in diameter, the central circle is about 80 Å in diameter and each of the 11 to 15 rods are 500 Å long. This structure does begin to disintegrate in EDTA concentrations much stronger than those needed to inhibit AF action (Henkart *et al.*, 1973). Similar sunburst configurations have been observed in *Microciona prolifera* AF preparations (Humphreys *et al.*, 1975).

The dimensions of the AF in these studies suggest it may be visible on MD cells and on CF cells to which AF has been added. Furthermore, if the sunburst configuration of AF is required for reaggregation at 5°C, it should be absent from CF cells. Two attempts to visualize AF *in situ* report that there is no apparent difference between cells in the intact sponge and CF cells (Revel and Goodenough, 1970; Anderson, 1975) and that 0.1 μm diameter fibers are present in both cases (Anderson, 1975). Unfortunately, neither of these brief reports contain micrographs. However, in the longer (Reed *et al.*, 1976) of two reports (Reed *et al.*, 1976; Greenberg and Pierce, 1973) on the effects of cytochalasins on reaggregation, scanning electron micrographs show that MD cells possess 0.1 μm diameter fibrils which are absent from MD cells that have been washed with calcium-free sea water. Cytochalasin-treated cells display a much denser network of these fibers than untreated cells. The authors suggest that overproduction of AF, which they equate with the fibrils, accounts for the inhibition of reaggregation they observe after cytochalasin treatment (Reed *et al.*, 1976). However, since there are a variety of fibrous extracellular materials present in sponges, including a collagen that reappears in the first hour after dissociation (Evans and Berquist, 1974), and since none of the studies just cited examined CF cells to which soluble AF had been restored, it is premature to equate the fibrils with AF.

Microciona parthena AF contains equal weights of protein and carbohydrate (Henkart *et al.*, 1973). *Microciona prolifera* AF also has been reported to consist of approximately equal weights of protein and carbohydrate (Humphreys *et al.*, 1975; Turner and Burger, 1973). Uronic acids comprise about 10% of the weight of *Microciona parthena* AF (Henkart *et al.*, 1973) and, according to one report (Turner and Burger, 1973), the same is true of *Microciona prolifera* AF, although an earlier study of the latter AF detected no uronic acid (Margoliash *et al.*, 1965). A more recent abstract has reported that further purification of *Microciona prolifera* AF than with the procedures previously employed causes the protein to carbohydrate ratio to rise from 1:1 to 2:1, with no uronic acid determinations reported (Humphreys *et al.*, 1975).

Light scattering and sedimentation both indicate that the molecular weight of *Microciona parthena* AF is approximately 2×10^7 daltons (Cauldwell *et al.*, 1973), which is the same value obtained by calculations based on its density in CsCl and size in electron micrographs (Henkart *et al.*, 1973). Viscosity determinations and the formation of a hypersharp boundary in the ultra-centrifuge both indicate that it is a fibrous molecule (Cauldwell *et al.*,1973), which agrees with its appearance in the electron microscope (Henkart *et al.*, 1973).

Both *Microciona prolifera* and *Microciona parthena* AFs undergo a decrease in molecular weight when they are treated with calcium-free solutions (Cauldwell *et al.*, 1973; Margoliash *et al.*, 1965). Treatment of the latter AF with successively greater concentrations of EDTA first causes the AF to lose its biological activity and then to irreversibly dissociate into a 2×10^5 dalton subunit (Cauldwell *et al.*, 1973) that is 1:1 protein: carbohydrate by weight. The 'arms' of the 'sunburst' seen in the electron microscope begin to unravel as the 2×10^5 dalton subunit forms (Cauldwell *et al.*, 1973). The intact AF binds irreversibly to anion exchangers (Henkart *et al.*, 1973) and has both high and low affinity Ca^{2+} binding sites (Cauldwell *et al.*, 1973).

Humphreys and his co-workers have suggested that their data (cited above, Cauldwell *et al.*, 1973; Henkart *et al.*, 1973) indicates that *Microciona parthena* AF is a 2×10^7 dalton complex composed of 2×10^5 dalton subunits whose association is stabilized by the presence of Ca^{2+} in the high affinity site. They propose that the low affinity calcium-binding sites are involved in cell adhesion, but they emphasize that the species specificity of *Microciona parthena* and *Microciona prolifera* AF action eliminates the possibility that these intercellular ligands act by simply trapping cells.

The observations of Burger and his co-workers (Turner *et al.*, 1974) provide information on the mechanism of interaction between *Microciona prolifera* AF and cells. Kuhns and Burger (1971) demonstrated that AF binds plant lectins and concluded that there are probably carbohydrate residues exposed on the AF surface. Burger, Lemon, and Radius (1971) and subsequently Turner and Burger (1973) determined that glucuronic acid and the glucuronic acid disaccharide, cellobiuronic acid, both inhibit AF-mediated reaggregation of *Microciona prolifera* CF cells, but not *Cliona celata* CF cells. This inhibition was unique to glucuronic and cellobiuronic acids and did not occur with several other monosaccharides, including galacturonic acid and glucose.

An impure mixture of glycosidases was found to destroy AF activity (Turner and Burger, 1973). Glycosidase treatment of AF in the presence of glucuronic acid did not affect AF activity but the glycosidase was effective if glucose or galacturonic acid were present during the incubation of the AF with the glycosidase (Turner and Burger, 1973). Subsequent analysis of glycosidase activity toward paranitrophenol glucuronate, galacturonate or glucose determined that the presence of glucuronic acid preferentially inhibited the glucuronidase activity of the glycosidase and that this glucuronidase activity was not inhibited by galacturonic acid or glucose (Turner, unpublished observations). Therefore, it appears that the ability of the crude glycosidase to destroy AF action is associated only with glucuronidase activity in the glycosidase. This result, in conjunction with the inhibition of AF-mediated reaggregation by glucuronic acid, is interpreted to indicate that glucuronic-like residues on the AF are recognized by some component of the *Microciona prolifera* CF cell surface (Turner and Burger, 1973; Turner *et al.*, 1974).

The cell surface component involved in this recognition event is released from the cells by osmotic shock (Weinbaum and Burger, 1973). When CF cells are treated with the appropriate low concentration of saline, the cells lose their ability to reaggregate with added AF. Incubating the 'shocked' cells with the low salt wash restores their ability to reaggregate with AF. The viability of the hypotonically shocked (HY) cells as determined by vital staining and the recovery of cells after the hypotonic shock treatment is greater than 90% (Jumblatt *et al.*, 1975).

Three important properties of the activity of the hypotonic shock supernatant, which has been termed baseplate (BP), indicate that it contains the specific binding site for AF:

(1) The baseplate inhibits AF activity toward 'unshocked' CF cells and this inhibition is sensitive to glucuronic acid (Weinbaum and Burger, 1973).
(2) Removing the baseplate from the hypotonically shocked cells (HY cells) causes them to lose their activity toward AF (Weinbaum and Burger, 1973).
(3) Adding BP back to HY cells restores their ability to undergo glucuronic acid-sensitive AF-mediated reaggregation (Weinbaum and Burger, 1973).

Since BP removal from and restoration to the HY cells controls their ability to undergo AF-mediated reaggregation, BP is more than just an AF inhibitor. Other AF inhibitors have been obtained by homogenization or trypsinization of CF cells (Humphreys, 1975). However, unlike BP, these inhibitors are apparently incapable of restoring the ability of the cells from which they are obtained to undergo AF-mediated reaggregation.

The existence of soluble BP permits the study of AF–BP interaction in the absence of intact cells. AF and BP have been covalently linked to separate sets of Sepharose beads by the cyanogen bromide procedure (Weinbaum and Burger, 1973; Cuatrecasas, 1970). Adding soluble AF to the BP beads caused the beads to agglutinate while the addition of soluble BP to the AF beads reversed the spontaneous aggregation of the AF beads. Both of these effects of the soluble component on the bead-bound component were sensitive to glucuronic acid, thereby strengthening the observation that BP–AF interaction is mediated by glucuronic acid.

The behavior of gluteraldehyde-fixed CF cells also supports a role of glucuronic acid in the BP–AF interaction (Jumblatt *et al.*, 1975). After gluteraldehyde fixation, CF cells undergo AF-mediated reaggregation which is inhibited by glucuronic acid or by preincubation of the AF with soluble BP. Furthermore, the aggregates formed by gluteraldehyde-fixed CF cells in the presence of AF are loose and irregular (Jumblatt *et al.*, 1975) and therefore resemble the loose aggregates formed either at low temperature (Humphreys, 1963) or by formalin-fixed cells (Moscona, 1963). It would appear that the fixed cells can form cell adhesions and that these adhesions occur by a glucuronic acid-sensitive interaction between BP and AF.

A variety of procedures have been unsuccessfully explored as potential methods for purifying AF by affinity chromatography on BP beads, and vice versa. However, some progress has been made in purifying BP activity by more conventional methods

(Jumblatt *et al.*, 1975). It should be emphasized that both the ability to inhibit AF-mediated CF cell reaggregation and the ability to restore aggregation to HY cells co-purify through differential centrifugation and Sephadex chromatography. The active column fraction has a molecular weight of approximately 45–60 000, is non-dialyzable, sensitive to 60°C for 10 min and inactivated by pronase. However, it is stable to extraction with lipid solvents, EDTA, lyophilization and a pH range of 3 to 12 (Weinbaum and Burger, 1973; Jumblatt *et al.*, 1975). These preliminary observations suggest that both the inhibitory activity and the ability to restore AF-mediated reaggregation to HY cells are associated with the same protein component of the BP preparation.

The ease of extraction of AF and of BP and the sensitivity of the BP–AF interaction to glucuronic acid all fluctuate from one season to the next. As Jumblatt *et al.* (1975) have indicated, the precise concentration of saline that extracts BP is variable and must be re-evaluated at the beginning of each season. The source of this variability is unclear, but not entirely unexpected, in view of the seasonal fluctuations in cell types present in the intact sponge (Simpson, 1968) and of the evidence that different cell types may have different roles in the reaggregation process (Galtsoff, 1923, 1925a, 1929; Johns *et al.*, 1971; Leith and Steinberg, 1972).

As we indicated in Section 6.4, Galtsoff (1923, 1925a, 1929) and Johns *et al.* (1971) found that archeocytes are essential for reaggregation in stationary cultures. Leith and Steinberg (1972) examined the ability of three different cell types from *Microciona prolifera* to reaggregate in the presence and in the absence of AF. These authors identify Nile Blue sulfate positive cells as archeocytes or grey cells, Nile Blue sulfate negative cells with red granules as collar cells and Nile Blue sulfate negative nucleolate cells with orange vacuoles were left unidentified. The Nile Blue sulfate positive cells went to the bottom of a 1 xg discontinuous sucrose/artificial sea water gradient, the collar cells stayed in the upper portion of the gradient and the unidentified cells appeared in the middle of the gradient. We feel that the unidentified cells are probably archeocytes, since they do not stain with Nile Blue sulfate, in agreement with the observation of Wilson and Penney (1930), and they do contain both nucleoli and orange vacuoles (Wilson and Penney, 1930; Ganguly, 1960). The cells that Leith and Steinberg identify as archeocytes or grey cells are probably the latter, since Wilson and Penney (1930) have identified grey cells as being Nile Blue sulfate positive. The only problem presented by this interpretation is that the grey cells, which are apparently intermediate in size between archeocytes and collar cells (Simpson, 1968), should occupy the middle, and not the bottom, of the gradients. However, this reversal of the position of the archeocytes and the grey cells might arise from differences in the response of the cells to the sucrose used to construct the gradient. If this tentative reinterpretation of the identity of the cells is accepted, the grey cells form small compact aggregates and the archeocytes form loose, flakey aggregates in the reports both of Leith and Steinberg (1972) and of Johns *et al.* (1971). Leith and Steinberg found that this difference in aggregate morphology existed between CF archeocytes and CF grey cells both before and after they were separated on the gradients. However, CF grey cells differed from MD grey cells because the latter were sometimes found on the surface of aggregates containing all three cell types, while CF grey cells were always in the interior of mixed aggregates. These observations suggest that cell type-specific differences exist in the association between AF and the cells, but they do not indicate the nature of the differences.

The results described thus far in this section indicate that *Microciona prolifera* AF has been studied extensively, but that there are still many unanswered questions about its mechanism of action. Like *Microciona prolifera* AF, *Haliclona occulata* AF is released during production of CF cells, it restores CF cell reaggregation to that of MD cells, restores temperature-insensitive cell adhesion formation to CF cells (Curtis, 1962) and the AF from *Haliclona variabilis* is active against fixed cells (Gasic and Galanti, 1966), as is *Microciona prolifera* AF (Moscona, 1968). Chemical analyses of *Haliclona occulata* AF indicate that it has a significantly smaller weight ratio of carbohydrate to protein than *Microciona prolifera* (Margoliash *et al.*, 1965; Turner and Burger, 1973) or *Microciona parthena* AF (Henkart *et al.*, 1973). The sensitivity of *Haliclona variabilis* AF to enzymatic and chemical treatment (Gasic and Galanti, 1966) indicates that the AF depends on the presence of disulfide bonds for its activity. The extent to which the analyses of *Haliclona variabilis* AF apply to *Haliclona occulata* AF is unclear, since it is not known whether or not these two species undergo species-specific reaggregation.

In the preceding section it was suggested that McClay's collecting aggregate studies (1971) show that *Haliclona viridis* and *Tedania ignis* form species-specific initial adhesions and that the other five species he examined form non-specific adhesions and then undergo species-specific cell sorting. McClay has detected AF activity in the dissociation supernatants from these sponges (McClay, 1974). The AFs apparently affect different cellular events than the AFs from *Microciona prolifera* and from *Haliclona occulata,* since they enhance aggregate fusion over an 8-hour period, rather than affecting initial cell adhesions. Since the AFs from *Haliclonia viridis* and *Tedania ignis* do not affect the formation of initial cell adhesions (McClay, 1974) and the cells of these two species do form species-specific initial adhesions, it appears that these two species' AFs are not involved in generating species—species adhesions.

In addition to detecting AF activity McClay (1974) found that some of his AF preparations actually inhibited the aggregation of some of the heterospecific cells. In one case, that of *Haliclona viridis,* the inhibitory activity was separable from the AF activity.

The role of inhibitory factors has been studied in greater detail in the species specificity of the second stage of reaggregation in the combination of *Geodia cydonium* and *Ircinia muscarum* (Müller *et al.*, 1976a). These results originated with the observation that the formation of large 'secondary' aggregates from small 'primary' aggregates during *Geodia cydonium* cell reaggregation is enhanced by a soluble factor (Muller and Zahn, 1973). Three important considerations prevent this observation from being analogous to the system established by Humphreys (1963):

(1) The *Geodia cydonium* cells are dissociated by trypsinization.
(2) The cells used in the AF assay are prepared by a different procedure than that used to prepare the AF itself.
(3) The AF clearly enhances only the formation of large aggregates from small aggregates, and does not enhance initial cell adhesion.

The first two points suggest that the reversible elution of AF from the cells in Humphreys' case is not occuring in the case of *Geodia cydonium*. The last point indicates that any specificity manifested by the AF does not involve the formation of specific cell adhesions. Futhermore, the AF-mediated formation of 'secondary' aggregates during *Geodia cydonium* reaggregation cannot be equated to the aggregate compaction Humphreys observed, since the former process is temperature-independent (Müller *et al.*, 1976b), while the latter is temperature-dependent (Humphreys, 1963). The *Geodia cydonium* AF apparently acts, at least in part, by enhancing the synthetic capabilities of the reaggregating cells (Müller *et al.*, 1976b) which may indicate the necessity for a period of 'repair' synthesis that may be required due to the use of trypsin to dissociate the sponge.

Geodia cydonium AF is sensitive to endopeptidases, heat inactivation and freezing and is insensitive to glycosidases, endopeptidases, DNase I and RNase (Müller and Zahn, 1973). Both AF-independent primary aggregate formation and AF-dependent secondary aggregate formation require Ca^{2+}, Sr^{2+}, or Ba^{2+} (Müller *et al.*, 1974). The AF activity is associated with a 'sunburst' configuration in the electron microscope in some reports (Müller and Zahn, 1973; Müller *et al.*, 1976a) and not in others (Müller *et al.*, 1974).

The ability of *Ircinia* cells to inhibit *Geodia cydonium* secondary aggregate formation depends upon the presence of sulfhydryl groups on both the AF and the inhibitor (Müller *et al.*, 1976b). This has led to the suggestion that the inhibitor acts by forming disulfide bridges between itself and the AF. The temperature independence of this process may indicate that it is non-enzymatic. The inhibitor is pronase-sensitive, glydosidase-insensitive and has a molecular weight of approximately 10 000. *Ircinia* mucoid cells contain most of the inhibitory activity and very little is found in the archeocytes. It appears that, as in the reports of Leith and Steinberg (1972) and Johns *et al.* (1971), the mucoid (grey ?) cells determine the species-specificity of sponge cell associations, even though the same type of associations are probably not being monitored in these three cases.

An AF 'receptor' that inhibits *Geodia cydonium* secondary reaggregation has been obtained from *Geodia cydonium* cells in a manner that did not leave the receptor-less cells intact. Thus, it is impossible to determine whether this inhibitor also has the second important property of a true receptor: the ability to restore AF reactivity to receptor-less cells, since a population of intact receptor-less cells was not produced. However, it was shown that some dissociated cells may be partially depleted of receptor. This was accomplished by demonstrating that trypsin-dissociated cells show enhanced AF-mediated secondary reaggregation in the presence of low concentrations of receptor and are inhibited by high receptor concentrations (Müller *et al.,* 1976c). The authors interpret the low receptor results as indicating that low concentrations of solute receptor attach to a low number of vacant receptor binding sites on the cells, thereby increasing the reactivity of the cells to the AF. The high concentration of soluble receptor is presumably binding to the soluble AF and therefore preventing its interaction with the cells.

It is interesting to note that *Microciona prolifera* AF appears to be approximately 50% carbohydrate (Henkart *et al.,* 1973; Turner *et al.,* 1974) and the baseplate is probably a protein (Weinbaum and Burger, 1973; Jumblatt *et al.,* 1975) while the *Geodia cydonium* AF is predominantly protein (Müller and Zahn, 1973; Müller *et al.,* 1976b) and its 'receptor' primarily carbohydrate (Müller *et al.,* 1976c). It is not known whether these reciprocal relationships between the carbohydrate: protein ratios of the two AFs and their respective putative 'receptors' bears any relationship to the fact that *Geodia cydonium* AF enhances aggregate to aggregate adhesions in contrast to *Microciona prolifera* AF which acts on cell to cell adhesion.

Van de Vyver and Curtis have found that the fresh water sponge *Ephydatia fluviatilis,* like *Microciona prolifera,* produces an AF that enhances the formation of specific cell–cell adhesions (Van de Vyver, 1975; Curtis and Van de Vyver, 1971). Van de Vyver distinguishes between strains of *Ephydatia fluviatilis* by the ability of the cells produced by germinating gemmules to co-operate and from one sponge (Van de Vyver, 1970; 1975). This co-operation is, by definition, strain-specific. The *Ephydatia fluviatilis* AF enhances strain-specific cell adhesions during the first 40 minutes of the assay (Van de Vyver, 1975; Curtis and Van de Vyver, 1971). This is in contrast to the non-specific cell reaggregation that occurs in the absence of the AF (Van de Vyver, 1975) (although in this case, the cells do sort out after the aggregates have formed). The apparent ability of this AF to produce a specificity that is not expressed by the cells alone may be due to the presence of an inhibitory factor in the AF preparation.

Curtis and Van de Vyver (1971) have reported that the AF preparation that enhances the aggregation of cells of the same strain inhibits reaggregation of cells from a different strain. This inhibitory activity is expressed in the presence of Ca^{2+}, while the enhancing activity is observed in the absence of Ca^{2+} (Van de Vyver, 1975; Curtis and Van de Vyver, 1971). We are unaware of any studies of the possibility that the inhibitory activity and the enhancing activity reside in different proteins in the AF preparation. It has been reported that the aggregation-promoting activity is found in a protein with a molecular weight of around 50 000 (Van de Vyver, 1975).

It is not clear that this AF preparation is obtained from the same cells used for the adhesion assay, which complicates any interpretation that the AF and cells are undergoing a reversible dissociation such as that which apparently exists for *Microciona prolifera.* In fact, in one report the cells are obtained from a gemmule sonicate (Curtis and Van de Vyver, 1971) which would seem to necessitate support-ing evidence before it could be concluded that the AF is being released from the cell surface. It has been reported that *Ephydatia fluviatilis* AF is removed from the medium during reaggregation (Van de Vyver, 1975), which indicates that the AF is binding to the cells.

Van de Vyver (1975) has reported that the interaction between two species of marine sponge is also mediated by inhibitory factors. When *Crambe crambe* and *Axinella polypoides* mechanically dissociated cells are mixed no aggregates form, nor are aggregates formed when cells of one of these species are mixed with a cell-free supernatant from the other species (Van de Vyver, 1975). This inhibitory activity is distinct from the hemagglutinating activity possessed by *Axinella polypoides* (Van de Vyver, 1975; Bretling and Kabat, 1976). The *Crambe crambe* inhibitor actually lysed half of the *Axinella polypoides* cells with which it was mixed. The *Axinella polypoides* inhibitor immobilized the *Crambe crambe* cells, which recalls the similar observations of Galtsoff (1925a, 1929) that, in some bispecific stationary cultures, one species' cells displayed decreased motility.

MacLennan and Dodd (1967) also obtained soluble factors from *Axinella* sp. by washing the cells in either calcium-free or complete sea water. These authors report that both types of washes aggregated cells from several other sponge species, but that only the cell-free supernatant obtained by calcium-free sea water washes was active against *Axinella* sp. cells.

MacLennan and Dodd attempted to obtain AF activity from several additional sponge species and succeeded in obtaining good AF preparations from *Hymeniacidon perleve* and feeble AF activity from *Halichondria panicea*. Neither of these AF's were species-specific, since they cross-reacted with each other and the *Hymeniacidon perleve* AF also cross-reacted with cells from *Ficulina (Suberites) ficus*, although the latter species did not produce an AF. It is not clear whether these AF's mediate cell-to-cell or aggregate-to-aggregate adhesions, since the AF activity was assayed after 16 hours of incubation (MacLennan and Dodd, 1967).

MacLennan (1963, 1969a and b, 1974) has studied the antigenic relationships between the surfaces of cells from various sponge species. MacLennan found a variable degree of cell surface antigen cross-reactivity that was related to the cross-reactivity between carbohydrate-containing macromolecules extracted from the intact sponges. The carbohydrate-containing materials were tested without success (MacLennan, 1969a) for their ability to both replace or inhibit the AF activity that had been shown (MacLennan and Dodd, 1967) to promote the amount of reaggregation present at 16 hours. It appears that sponge cell surface antigens are due, at least in part, to their carbohydrates, but that these carbohydrates are not involved in determining the extent of reaggregation present after 16 hours of incubation. It is not clear that these antigenic, or other non-antigenic cell surface carbohydrates are unimportant in the initial stages of cell adhesion.

MacLennan (1974) has recently turned his attention to sponge extracts that have hemagglutinating activity. The *Axinella* sp. hemagglutinin is specific for galactose or galactosamine residues (MacLennan, 1974; Bretling and Kabat, 1976), but these monosaccharides do not inhibit *Axinella* sp. cell agglutination by the hemagglutinin nor the interaction between the hemagglutinin and *Axinella* sp.

carbohydrates that are immunologically related to the *Axinella* sp. cell surface. Hemagglutinins have also been extracted from *Cliona celata* (MacLennan, 1974) and *Aaptos papillata* (Bretling *et al.*, 1976), thus adding sponges to the list of plants and invertebrates that contain lectins.

The studies outlined in this section have described three types of soluble molecules from sponges that affect cell adhesions: aggregation factors, inhibitory factors and hemagglutinins. The sponge hemagglutinins appear to share with other plant and invertebrate lectins the ability to agglutinate erythrocytes by binding to cell surface carbohydrates (Nicholson, 1974). The sponge lectins have not been shown to have a role in a sponge cell reaggregation since they can be separated from AF activity (Van de Vyver, 1975).

The aggregation factors described above have all been assayed by determining their ability to enhance reaggregation. The methods of measuring reaggregation differ in each set of experiments, but can be considered in a simplistic manner to detect:

(1) Cell–cell adhesions in the studies of Humphreys (1963), Burger (Turner *et al.*, 1974) and Curtis and Van de Vyver (1971).
(2) Some combination of cell–cell, cell–aggregate and aggregate–aggregate adhesions in the reports by McClay (1974) and by MacLennan and Dodd (1967).
(3) Obvious aggregate–aggregate adhesions in the case of Müller and co-workers (Müller and Zahn, 1973).

These three classes of assays do not provide any direct information concerning the mechanism of AF action. However, it is probable that some cell adhesions have already occurred before the AFs described in the last two cases have begun to act, since these AFs affect, at least in part, events occurring after aggregate formation begins. If this conclusion is correct, it implies that in these studies cell-to-cell adhesions are not totally dependent on AF activity.

A second point to be emphasized is that in order to conclude that the mechanism of cell-to-cell adhesion is being examined it is necessary to demonstrate not only that the AF enhances reaggregation but also that the AF restores the reaggregation capabilities of the cells from which it has been removed. *Microciona prolifera* AF enhances the reaggregation of MD cells (Weinbaum and Burger, 1973), but this result alone does not define the AF as a component of a cell-to-cell adhesion mechanism. This definition requires that the AF be shown to restore to the CF cells from which it was removed the ability to reaggregate in a manner that is qualitatively identical to and quantitatively very similar to the reaggregation of MD cells without added AF. All the essential components of this system — MD cells whose reaggregation can be qualitatively and quantitatively defined, CF cells that have reduced reaggregation relative to MD cells and are also the source of AF, and the AF which restores to the CF cells the reaggregation capabilities of the MD cells — must be present to permit the conclusion that the AF has a role in cell-to-cell adhesion. Much of the research discussed here does not analyze one of the essential cell populations which is necessary to draw this conclusion. This analysis obviously suffers from an inability to provide any insight into the importance or mechanism of action of the AFs other than that from *Microciona prolifera*. It is apparent that the other AFs are capable of enhancing reaggregation, but it has yet to be proven that they do so because they are components of the cell-to-cell adhesion system. It is also obvious that, for reasons that are not understood, the properties of *Microciona prolifera* cell—cell adhesions are amenable to this direct approach. Continued study should elucidate both the properties of the *Microciona prolifera* system that make it unique and the precise role of the other AFs.

The species specificity of *Microciona prolifera* AF activity suggests that it is involved in cell recognition as well as cell adhesion. As pointed out in previous sections, this species-specificity is also observed in the behavior of *Microciona prolifera* cells and this specificity of sponge cell adhesion is not common to all species. The species-specific action of other AFs is not necessarily expressed at the level of cell—cell adhesions. McClay (1974) has reported 5 cases of species-specific AFs but, as suggested above, it is not clear that the AFs necessarily act at the level of cell adhesions, since the assay used could also be measuring cell-to-aggregate and aggregate-to-aggregate adhesions. While the possibility still exists that cell adhesions are being assayed, it is also possible that the formation of secondary aggregates, such as those detected by Müller and Zahn (1973), is being monitored and that this phase of reaggregation is displaying a species-specific response to the AFs.

McClay (1974) has also demonstrated the presence of heterospecific inhibitory agents in a number of the crude AF preparations and has separated the inhibitory activity from the AF in one case. Inhibitory activity in AF preparations has been reported to produce strain-specific cell adhesions by Curtis and Van de Vyver (1971) and to generate species-specific secondary aggregate formation by Müller *et al.* (1976a). All attempts to distinguish AF and inhibitory activity have been successful (Van de Vyver, 1975; McClay, 1974; Muller *et al.*, 1976a), which suggests that different molecular components of the crude supernatants are responsible for the AF and the inhibitory activity.

6.7 CELL RECOGNITION AND SPONGE CELL BIOLOGY

The comments in this closing section present some observations on five aspects of sponge biology as they may affect the interpretation of studies of cell adhesion. The aim of this section is to try to place the *in vitro* experiments described in the previous sections in an *in vivo* framework. It will become obvious that the relationship of *in vitro* reaggregation studies to sponge cell associations *in vivo* requires further study.

6.7.1 Fluctuations in cell populations

One of the first facts to be considered is that the sponges that serve as a source of cells for the *in vitro* studies undergo seasonal changes *in vivo* that may affect the *in vitro* observations. Van de Vyver (1975) has demonstrated that cells from young sponges reaggregate better than those from older sponges, and suggested that this may be due to the presence of fewer archeocytes in the older sponges. Archeocytes have a central role in reaggregation (Van de Vyver, 1975; Johns *et al.,* 1971; Leith and Steinberg, 1972) and are the probable source of eggs, sperm and embryos *in vivo* (Simpson, 1968). Any seasonal fluctuations in the proportion of archeocyte present in a sponge which may occur as a consequence of the sponge's reproductive cycle will probably influence the outcome of *in vitro* studies. The variability in the reproductive cycle that can exist is illustrated by Fell's observation (1974) that *Haliclona loosanoffi* in one location on Long Island Sound contains reproductive elements in the early summer, while the same species at a different Long Island Sound location has reproductive elements present late in the summer. We have already pointed out that in addition to seasonal fluctuations, the nutrition of the sponge may affect the cell types present, since starvation triggers the transformation of archeocytes into choanocytes (Agrell, 1951). When these fluctuations in the cellular compostion of the intact sponge are considered in conjunction with the variable contributions that different cell types may make both to the composition of cell suspensions (Wilson and Penney, 1930; Galtsoff, 1923; 1925a; deLaubenfels, 1934; Borojevic and Levi, 1964; Mookerjee and Ganguly, 1961) and to the *in vitro* adhesion process (Johns *et al.,* 1971; Leith and Steinberg, 1972) it becomes apparent that definitive interpretations of the *in vitro* adhesion results depend on determinating the cell types present in the cell suspensions used for the adhesion assays.

6.7.2 Surface coats

In addition to the cellular variations just discussed, a sponge surface coat that lies external to the pinacocytes shows seasonal fluctuations, since it is present in the fall and absent from sponges collected in the spring (Bagby, 1970). This coat is composed of an amorphous material in which fibers are frequently observed (Bagby, 1970). Two additional sites at which surface coats have been found are on the surface of aggregates formed in stationary culture (Bagby, 1972) and on the surface of cells in the intact sponge (Garrone *et al.,* 1971). The cell coat has been demonstrated to contain both glycoproteins and acidic mucopolysaccharides (Garrone *et al.,* 1971), but analyses of the aggregate surface coat have not been reported. Therefore, the precise nature of these surface coats, as well as their relationship to each other and to the AFs, inhibitory factors, hemagglutinins or antigenic carbohydrates discussed above (Section 6.6) has yet to be determined. However, the production of the aggregate coat between 12 and 24 hours of stationary culture (Bagby, 1972), suggests that it is present at the time that aggregate-to-aggregate interactions occur in rotation-mediated reaggregation. A detailed examination that related aggregate fusion to the appearance of the aggregate surface coat would indicate whether the aggregate coat forms in rotation as well as in stationary cultures and, if it does, if it has a role in aggregate-to-aggregate interactions. The site of initial contact between two out-growths or two gemmules may be the sponge surface coat, since it is external to the pinacocytes. The sponge surface coat therefore warrants further study since it may be one component of gemmule to gemmule and outgrowth to outgrowth interaction mechanisms.

6.7.3 Extracellular matrix

The extracellular matrix of the intact sponge contains the surface coats described above, two distinct collagenous proteins (Gross *et al.*, 1956; Garrone *et al.*, 1975; Cowden and Harrison, 1976; Garrone and Pottu, 1973; Katzman *et al.*, 1972; Junqua *et al.*, 1974) and an amorphous material (Gross *et al.*, 1956; Garrone and Pottu, 1973; Junqua *et al.*, 1974; 1975; Stempien, 1966; Cowden, 1970; Smith and Lauritis, 1969).

These collagenous components include a molecule very similar to vertebrate collagen, spongin A, and a molecule unique to sponge, spongin B, which are both synthesized by spongocytes that develop from archeocytes (Garrone and Pottu, 1973). The spongins both have a large quantity of hydroxyproline and also contain carbohydrate, including hexosamine, neutral hexose and uronic acid (Gross *et al.*, 1956). Spongins A and B also have a relatively high ratio of hydroxylysine to lysine and a large percentage of the hydroxylysine is glycosylated. These last two properties cause the spongins to resemble other invertebrate collagens and the collagen that is thought to be present in vertebrate basement membranes (Katsman *et al.*, 1972; Junqua *et al.*, 1974). The hydroxyproline and hydroxylysine content of the spongins and their supramolecular structure distinguishes them from all AFs that have been described, suggesting that the collagen-like material that is released in the first hour of reaggregation (Evans and Berquist, 1974) is not directly involved in reaggregation.

In addition to the sponge surface coat, the cell coat and spongins A and B, the sponge extracellular matrix also consists of an 'amorphous material' (Gross *et al.*, 1956) that contains glycoproteins (Garrone and Pottu, 1973). This ground substance contains a variety of carbohydrates (Junqua *et al.*, 1974) and cytochemical analyses of sectioned sponges demonstrate the presence of both sulfate and carboxyl groups (Stempien, 1966; Cowden, 1970). However, further studies are necessary to define the relationship of this amorphous material to the ground substance present in other organisms. The cellular source of this amorphous material is not clear, since various reports using a number of sponge species suggest that it is produced by archeocytes (Galtsoff, 1925a and b), globoferous cells (Simpson, 1963; 1968) or rhabdiferous cells (Simpson, 1963; Smith and Lauritis, 1969).

We have discussed observations that implicate archeocytes and mucoid (grey) cells in reaggregation (Johns *et al.*, 1971; Leith and Steinberg, 1972). If the archeocytes are among the cells that determine adhesive specificity *in vitro* (Galtsoff, 1929) and also produce some component of the ground substance *in vivo*, it is possible that *in vitro* cell—cell adhesiveness arises from interactions between cells and components of the ground substance. Even those AFs that do not affect cell—cell adhesions may occur in the ground substance *in vivo*. A number of the AFs we have discussed enhance aggregate size after cell—cell adhesions occur. In these cases, it is possible that the AF acts *in vitro* by contributing to attempts to reconstruct the extracellular matrix during the course of the reaggregation assay. AF enhancement of aggregate size could be a consequence of more extensive reconstruction of the extracellular matrix in the presence of AF than in its absence. These considerations suggest that some or all of the AFs may be found in the ground substance *in vivo*. Humphreys' observations of the structural and chemical similarities between *Microciona parthena* AF and the proteoglycans (Henkart *et al.*, 1973) of the extracellular matrix support this suggestion. Thus, it may not be necessary that all AFs be derived directly from the cell membrane in order to affect reaggreation nor is it necessary that they directly affect cell-to-cell adhesion.

6.7.4 Sponge cell junctions *in vivo*

Cell—cell adhesions in intact sponges have been examined ultrastructurally by electron microscopic studies of cell junctions (Reed *et al.*, 1976., Bagby, 1970; Jones, 1966; Ledger, 1975). The only report of specialized cell junctions in intact sponges has found septate desmosomes between sclerocytes and suggests that these junctions could form a seal between the spicule that the sclerocyte secretes and the ground substance of the sponge (Ledger, 1975). However it should be recalled that, in spite of the absence of gap junctions in ultrastructural studies of the intact sponge, two sponge cells can become electrically coupled (Lowenstein, 1967), which is a process that is generally thought to require gap junctions. Specialized junctions have not been observed between choanocytes, pinacocytes and cells from the mesohyl (Reed *et al.*, 1976; Bagby, 1970; Jones, 1966; Ledger, 1975). The latter category includes archeocytes, grey cells, rhabdiferous cells and globoferous cells. The membranes of all these cells are separated by a 10—30 nm gap (Bagby, 1970; Jones, 1966; Ledger, 1975) that contains a slightly dense, amorphous material which is apparently continuous with both the mesohyl ground substance and with the mucopolysaccharide cell coat. This amorphous material provides visual evidence for continuity between cell adhesive sites and the non-collagenous components of the sponge extracellular matrix. The relationship of the AFs to this amorphous material or other cell junction components has not been investigated. One possible relationship is that *Microciona prolifera* AF may be all or a portion of the amorphous material in the 10—30 nm gap between the cells. The *Microciona prolifera* BP, based on its release from the cells by hypotonic shock, may be an extrinsic membrane protein that serves as the AF binding site.

Continued investigation of the molecular interactions in this cell adhesion system will be of particular interest to membrane biologists, due to the unique fatty acid composition of *Microciona prolifera*. A significant proportion of the phospholipid fatty acids in this sponge are between 24 and 28 carbons long, instead of the usual 14 to 22 carbon chain length and contain an isolated double bond (Jefferts *et al.*, 1974;

Litchfield and Morales, 1976; Morales and Litchfield, 1976). Although it has yet to be demonstrated that these unique fatty acids are associated with the plasma membrane, the potential exists for studying a biological membrane with a unique fatty acid content and the effect of these unique fatty acids on cell recognition and adhesion.

6.7.5 Sponge 'Immune system'

The heteroagglutinins that were first described by Galtsoff (1929) resemble the AFs in that they differentiate 'self' from 'non-self'. The heteroagglutinins differ from the AFs in that they agglutinate cells recognized as 'non-self' while the AFs enhance reaggregation of 'self' cells. Galtsoff (1925a) provided the first evidence for inhibitory factors which also discriminate 'self' and 'non-self', but act by inhibiting the movement (Galtsoff, 1925a), decreasing the reaggregation (Curtis and Vande de Vyver, 1971) or actually lysing (Van de Vyver, 1975) heterologous cells. The role of the hetero-agglutinins, like that of the hemagglutinins, is not clear at the present. In contrast, the action of the inhibitory factors *in vitro* may have a direct parallel *in vivo*, as suggested by the studies of *Ephydatia fluviatilis* gemmules and cell suspensions (Van de Vyver, 1970; 1975; Curtis and Van de Vyver, 1971). Finally, the lytic agents and cell movement inhibitors may cause the necrotic zone that occurs at the interface between two outgrowths or grafts that do not fuse.

The recognition and destruction of heterologous organisms such as that just outlined is a simple definition of an immune system, and this definition indicates that these inhibitory agents deserve further investigation. Studies of their mechanism of action and comparisons to the vertebrate immune system should lead to increased understanding of the evolution and mechanism of action of vertebrate immune systems. Finally, the degree of specificity of the sponge inhibitory factors requires clarification: does this primitive system operate at the level of the individual, as indicated by the lack of fusion of individual *Microciona prolifera* outgrowths (Simpson, 1973), or is strain specificity, such as that found in *Ephydatia fluviatilis,* the maximum discrimination that can be attained?

6.8 SUMMARY

Cell recognition in sponges is a process that occurs *in vivo* at a strain-specific level and perhaps at the level of the individual sponge. *In vitro* analysis of the cellular and biochemical basis of this recognition have concentrated on species-specific cell recognition. Evidence has been provided that preferential formation of homologous cell adhesions, non-specific adhesion followed by cell sorting and inhibition of the activities of heterologous cells are all used to generate homospecific cell populations *in vitro*. Distinctions between those reaggregation studies that indicate that cell recognition precedes or accompanies cell adhesion and those in which cell adhesion clearly precedes cell sorting are necessary in order to distinguish the first two possibilities. These distinctions are only possible when reaggregation is monitored both early in the assay, while cell adhesions are forming, and late in the assay when cell sorting is occuring. Biochemical analyses of cell recognition involve the production of aggregation factors. These soluble molecules must be shown both to enhance reaggregation when added to cells and to reduce the reaggregation of the cells from which they have been removed in order to conclude that they affect cell–cell adhesions. In most cases, only enhanced reaggregation has been demonstrated, which allows several alternative interpretations. In the most intensively studied case of sponge cell adhesion, it has been shown that cells that lack the aggregation factor have lost some reaggregation capabilities, that addition of the factor restores these capabilities and that these effects are species-specific. Biochemical analyses of this system have defined the presence of carbohydrate in the factor, the existence of a soluble molecule with the properties of a factor binding site and suggest that the binding site recognizes a particular class of carbohydrate on the factor. The meaning of the *in vitro* results to the *in vivo* organization of the intact sponge is explored. Emphasis is placed on the need to define the location and role of the AFs in the intact sponge and the role of specific types of sponge cells in the *in vitro* studies of sponge cell adhesion.

REFERENCES

Agrell, I. (1951), *Arkiv. für Zool.*, **2**, 519–523.

Anderson, J.M. (1975), *Biol. Bull.*, **149**, 419.

Bagby, R.M. (1970), *Zeit. Zellforsch.*, **105**, 579–594.

Bagby, R.M. (1972), *J. exp. Zool.*, **180**, 217–244.

Bhakdi, S., Knufermann, H. and Wallach, D.F.H. (1974), *Biochim. biophys. Acta*, **345**, 448–457.

Borojevic, R. and Levi, C. (1964), *Z. Zellforsch.*, **64**, 708–725.

Bretling, H. and Kabat, E.A. (1976), *Biochemistry*, **15**, 3228–3236.

Bretling, H., Kabat, E.A., Liao, J. and Pereira, M.E.A. (1976), *Biochemistry*, **15**, 5029.

Burger, M.M., Lemon, S.M. and Radius, R. (1971), *Biol. Bull.*, **141**, 380.

Cauldwell, C.B., Henkart, P. and Humphreys, T. (1973), *Biochemistry*, **12**, 3051–3055.

Cowden, R.R. (1970), *Zeit. Mikroscop. Anat. Forschung*, **82**, 557–569.

Cowden, R.R. and Harrison. F.W. (1976), *Aspects of Sponge Biology*, Academic Press, New York, pp. 69–82.

Cuatrecasas, P. (1970), *J. biol. Chem.*, **245**, 3059–3065.

Curtis, A.S.G. (1962), *Nature*, **196**, 245–248.

Curtis, A.S.G. (1970a), *Nature*, **226**, 260–261.

Curtis, A.S.G. (1970b), *Symp. Zool. Soc. (Lond.)*, **25**, 335–352.

Curtis, A.S.G. and Van de Vyver, G. (1971), *J. Embryol. exp. Morph.*, **26**, 295–312.

deLaubenfels, M.W. (1927), *Carnegie Inst. Wash., Yearbook*, **26**, 219–222.

deLaubenfels, M.W. (1928), *J. Elisha Mitchell Sci. Soc.,* **44**, 82−86.

deLaubenfels, M.W. (1934), *Papers Tortugas Lab.,* **28**, 38−68.

Evans, C.W. and Berquist, P.R., (1974), *J. Microscopie.,* **21**, 185−188.

Fell, P.E. (1974), *Biol. Bull.,* **147**, 333−351.

Galtsoff, P.S. (1923), *Biol. Bull.,* **45**, 153−161.

Galtsoff, P.S. (1925a), *J. exp. Zool.,* **42**, 183−222.

Galtsoff, P.S. (1925b), *J. exp. Zool.,* **42**, 223−250.

Galtsoff, P.S. (1929), *Biol. Bull.,* **57**, 250−260.

Ganguly, B. (1960), *Wilhelm Roux' Arch. Entwicklungsmech. Organismen,* **152**, 22−34.

Garrone, R., Huc, A. and Junqua, S. (1975), *J. Ultrastruct. Res.,* **52**, 261−275.

Garrone, R. and Pottu, J. (1973), *J. Submicros. Cytol.,* **5**, 199−218.

Garrone, R., Thiney, Y. and Pavans de Cecalty, M. (1971), *Experentia,* **27**, 1324−1326.

Gasic, G.J. and Galanti, N.L. (1966), *Science,* **151**, 203−235.

Gerisch, G. (1960), *Wilhelm Roux' Arch. Entwicklungsmech. Organismen,* **152**, 632−654.

Greenberg, M.J. and Pierce, S.K., Jr. (1973), *Am. Zool.,* **13**, 1336−1337.

Gross, J., Sokal, Z. and Rouguie, M. (1956), *J. Histochem. Cytochem.,* **4**, 227−246.

Harrison, J.W. (1974), *J. Morph.,* **144**, 185−194.

Henkart, P., Humphreys, S. and Humphreys, T. (1973), *Biochemistry,* **12**, 3045− 3050.

Humphreys, T. (1963), *Dev. Biol.,* **8**, 27−49.

Humphreys, T. (1965), *J. exp. Zool.,* **160**, 235−240.

Humphreys, T. (1969), *Symp. Zool. Soc., (Lond.),* **25**, 325−334.

Humphreys, T. (1970), *Nature,* **228**, 685−686.

Humphreys, T. (1975), *Cellular Membrane and Tumor Cell Behavior*: Twenty-Eighth Annual Symposium on Fundamental Cancer Research. The Williams and Wilkens Co., Baltimore, MD., pp. 173−192.

Humphreys, T., Yonemoto, W., Humphreys, S. and Anderson, D. (1975), *Biol. Bull.,* **149**, 430.

Jefferts, E., Morales, R.W. and Litchfield, C. (1974), *Lipids,* **9**, 244−247.

Johns, H.A., Campo, M.S., MacKenzie, A.M. and Kemp, R.B. (1971), *Nature, New Biol.,* **230**, 126−128.

Jones, W.C. (1966), *J. Royal Micros. Soc.,* **85**, 53−62.

Jumblatt, J.E., Weinbaum, G., Turner, R., Ballmer, K. and Burger, M.M. (1975), *NATO Adv. Stud. Inst.,* **75**, 2.

Junqua, S., Fayolle, J. and Robert, L. (1975), *Comp. Biochem. Physiol.,* **50B**, 305−309.

Junqua, S., Robert, L., Garrone, R., Pavans de Cecatty, M., and Vacelet, J. (1974), *Connect. Tiss. Res.,* **2**, 193−203.

Katzman, R.L., Halford, M.H., Reinhold, N.V. and Jeanloz, R.W. (1972), *Biochemistry,* **11**, 1161−1167.

Kuhns, W.J. and Burger, M.M. (1971), *Biol. Bull.,* **141**, 393−394.

Ledger, P.W. (1975), *Tissue and Cell,* **7**, 13−18.

Leith, A.G. and Steinberg, M.S. (1972), *Biol. Bull.,* **143**, 468.

Levi, C. (1956), *Archs. Zool. exp. gen.,* **93**, 1–181.

Litchfield, C. and Morales, R.W. (1976), *Aspects of Sponge Biology,* Academic Press, New York, pp. 183–200.

Loewenstein, W.R. (1967), *Dev. Biol.,* **15**, 503–520.

McClay, D.R. (1971), *Biol. Bull.,* **141**, 319–330.

McClay, D.R. (1974), *J. exp. Zool.,* **188**, 89–102.

MacLennan, A.P. (1963), *Biochemistry,* **89**, 99.

MacLennan, A.P. (1969a), *Symp. Zool. Soc. (Lond.),* **25**, 299–324.

MacLennan, A.P. (1969b), *J. exp. Zool.,* **172**, 253–266.

MacLennan, A.P. (1974), *Arch. Biol.,* **85**, 53–90.

MacLennan, A.P. and Dodd, R.Y. (1967), *J. Embryol. exp. Morph.,* **17**, 473–480.

Margoliash, E., Schenck, J.R., Hargie, M.P., Burokas, S., Richter, W.R., Barlow, G.H. and Moscona, A.A. (1965), *Biochem. biophys. Res. Comm.* **20**, 383–388.

Mookerjee, S. and Ganguly, B. (1964), *Wilhelm Roux' Arch. Entwicklungsmech. Organismen,* **155**, 525–534.

Morales, R.W. and Litchfield, C. (1976), *Biochim. biophys. Acta.,* **431**, 206–216.

Moscona, A.A. (1961), *Exp. Cell Res.,* **22**, 455–471.

Moscona, A.A. (1963), *Proc. natn. Acad. Sci., U.S.A.,* **49**, 742–747.

Moscona, A.A. (1968), *Dev. Biol.,* **18**, 250–277.

Muller, W.E.G., Muller, I., Kurelec, B. and Zahn, R.K. (1976a), *Exp. Cell Res.,* **98**, 31–40.

Muller, W.E.G., Muller, I. and Zahn, R.K. (1974), *Experentia,* **30**, 899–902.

Muller, W.E.G., Muller, I. and Zahn, R.K. (1976b), *Biochim. biophys. Acta,* **418**, 217–225.

Muller, W.E.G., Muller, I., Zahn, R.K. and Kurelec, B. (1976c), *J. Cell Sci.,* **21**, 227–241.

Muller, W.E.G. and Zahn, R.K. (1973), *Exp. Cell Res.,* **80**, 95–104.

Nicolson, G.L. (1974), *Int. Rev. Cytol.,* **39**, 89–109.

Paris, J. (1960), Theses. Causse, Graille, Gastelnau, impremeurs, Montpellier, 1–74.

Reed, C., Greenberg, M.J. and Pierce, S.K., Jr. (1976), *Aspects of Sponge Biology,* Acadmic Press, New York, pp. 153–169.

Revel, J-P. and Goodenough, D.A. (1970), *Chemistry and Molecular Biology of the Intercellular Matrix,* Academic Press, New York, pp. 1361–1380.

Roth, S.A. and Weston, J.A. (1967), *Proc. natn. Acad. Sci., U.S.A.,* **58**, 974–980.

Sara, M., Liaci, L. and Melone, N. (1966), *Nature,* **210**, 1167–1168.

Simpson, T.L. (1963), *J. exp. Zool.,* **154**, 135–147.

Simpson, T.L. (1968), *J. exp. Mar. Biol. Ecol.,* **2**, 252–277.

Simpson, T.L. (1973), Colonialism among the Porifera. In: *Animal Colonies.* Dowden, Hutchinson and Ross, Stroudsburg, Pa., pp. 547–565.

Sindelar, W.F. and Burnett, A.L. (1967), *J. gen. Physiol.,* **50**, 1089–1090.

Smith, D.G. (1976), *Trans. Am. Micros. Soc.,* **95**, 235–236.

Smith, V.E. and Lauritis, J.A. (1969), *J. Microscopie,* **8**, 179–188.

Steinberg, M.S. (1970), *J. exp. Zool.,* **173**, 395–434.

Stempien, M.F. Jr., (1966), *Am. Zool.,* **6**, 363.

Turner, R.S. and Burger, M.M. (1973), *Nature,* **244**, 509–510.

Turner, R.S., Weinbaum, G., Kuhns, W.J. and Burger, M.M. (1974), *Arch. Biol.,* **85**, 35–51.

Van de Vyver, G. (1970), *Ann. Embryol. Morphogen.,* **3**, 251–252.

Van de Vyver, G. (1975), *Current Topics of Developmental Biology,* Vol. X, Academic Press, New York, pp. 123–159.

Weinbaum, G. and Burger, M.M. (1973), *Nature,* **244**, 510–512.

Wilson, H.V. (1907), *J. exp. Zool.,* **5**, 245–258.

Wilson, H.V. (1910), *Bull. Bur. Fish., Wash.,* **30**, 1–30.

Wilson, H.V. and Penney, J.T. (1930), *J. exp. Zool.,* **56**, 73–148.

7   Cell Adhesion in the Cellular Slime Molds

S T E V E N   D.   R O S E N   and   S A M U E L   H.   B A R O N D E S

Acknowledgements

Dr Rosen's research is supported by NIGMS grants RO1 GM23547 and RCOA KO4 GMOO322. Dr Barondes' research is supported by NIMH 18282 and a grant from the McKnight Foundation.

7.1 INTRODUCTION

The cellular slime molds are widely considered to be well-suited to the investigation of many fundamental problems in differentiation and morphogenesis. A faith in conservation of basic mechanisms throughout evolution encourages the hope that what is revealed about developmental processes in this lower eukaryote may be relevant for the understanding of related phenomena in other systems. Several recent reviews (Bonner, 1967, 1971; Gerisch, 1968; Newell, 1971; Garrod, 1974; Olive, 1974; Killick and Wright, 1974; Loomis, 1975; Jacobson and Lodish, 1975; Sussman and Brackenbury, 1976) have surveyed the considerable progress made to date in understanding the molecular and developmental biology of this organism. In this review, we shall emphasize the problem of cell adhesion in the cellular slime molds. We shall first review biological studies of cell aggregation and adhesion and then biophysical, biochemical, and immunological studies of the cell surface, which relate to the mechanism of cell adhesion.

7.2 THE ORGANISM

7.2.1 General description

The discussion of cellular slime molds will be limited to the genera *Dictyostelium* and *Polysphondylium*, which contain 20 and 3 identified species, respectively (Raper, 1973). These two genera comprise the family Dictyosteliaceae, which is in the order Acrasiales and the kingdom Protista.

A general overview of the life cycle is given below. A fuller description can be found in the reviews by Raper (1941) and Bonner (1967). A characteristic feature of these organisms is that they possess distinct non-social and social phases in their life cycle. In the non-social stage, which is the growth or vegetative stage, small solitary amoebae feed on bacteria by phagocytosis and divide by fission every 3–4 hours. Essentially, each amoeba behaves independently showing no consistent orientation with respect to its neighbors and forming no stable associations. When the food supply is depleted the growth phase terminates and, after an interphase of several hours, the social stage, involving a series of complex cellular interactions, commences. Guided by a chemotactic system, the amoebae align and move over large distances toward discrete aggregation centers. With the notable exception of *Dictyostelium minutum,* aggregation involves the formation of definite and conspicuous streams converging toward a center in a radial, wheel-like pattern or in a whirlpool fashion. Within the streams, the cells are closely adherent with their long axes predominantly aligned in the direction of movement. At each center, the

235

cells assemble into a peg-like multicellular aggregate called a pseudoplasmodium or slug which is surrounded by a slime sheath. These multicellular units may contain many thousands of cells; but they may be smaller if there is a low density of amoebae at the time of starvation. The subsequent development of the pseudoplasmodium varies with species. In some, it may migrate in response to environmental stimuli before entry into the program for fruiting body construction. Other species begin fruiting immediately at the original site of aggregation. The fruiting body which marks the endpoint of differentiation consists of two distinct differentiated cell types: the vacuolated structural cells of the stalk and the spore cells, responsible for continuation of the line. In general, stalk cells arise from the anterior portion of the pseudoplasmodium, whereas spore cells, which are in the majority, derive from the posterior section. Depending on species, the fruiting body consists of one or more spore caps borne on an unbranched or branched stalk, consisting of dead cells encased in a cellulose sheath. In some instances, two species are distinguished principally by the pigmentation of their spore caps. For example, *Dictyostelium purpureum* closely resembles *Dictyostelium mucoroides* in overall morphology, but differs in having dark purple pigmentation of its spore mass. Similarly, the violet spore mass of *Polysphondylium violaceum* distinguishes it from *Polysphondylium pallidum.*

The life cycle of the most studied species, *Dictyostelium discoideum,* is depicted in Fig. 7.1.

7.2.2 Chemotaxis

(a) *Dictyostelium discoideum*

Early work in this species by Runyon (1942), Bonner (1947) and Shaffer (1953) indicated that amoebae were guided to aggregation centers by a diffusable chemotactic substance termed acrasin. Recent investigations (Konijn *et al.,* 1967; Bonner *et al.,* 1969; Robertson *et al.,* 1972; Shaffer, 1972) have established that the acrasin is 3′, 5′ cAMP. As mentioned above, aggregation in *D. discoideum* does not consist of simple radial movement of individual amoebae toward a center. Rather, cells join with neighbors to form streams, which converge in larger streams, eventually merging at the center. Temporally, the cells move inward in a series of pulses which start at the center and propagate outward. This complex spatiotemporal patterns of aggregation is based on a chemotactic relay system (Shaffer, 1957a) involving several competencies which develop in a population of differentiating amoebae (reviewed by Robertson and Cohen, 1972). In a field of amoebae deprived of food for several hours, aggregation is initiated by the periodic secretion of acrasin by a few randomly distributed cells (Cohen and Robertson, 1971). A neighboring cell receiving a supra-threshold concentration of cAMP initiates an all-or-none step of movement toward the source (Konijn *et al.,* 1967; Cohen and Robertson, 1971) and releases a pulse of cAMP of its own (Robertson *et al.,* 1972; Shaffer, 1972). Relaying shows refractoriness in that the cells cannot be stimulated to relay again for several minutes

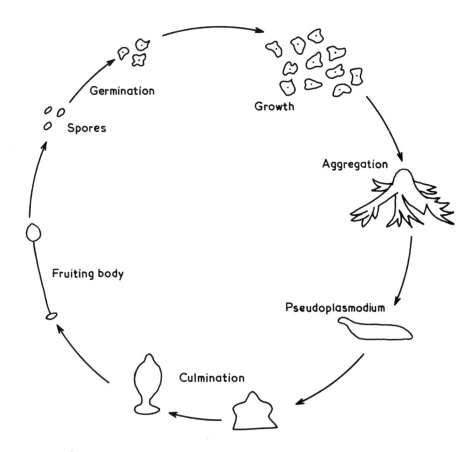

Fig. 7.1 Life Cycle of *Dictyostelium discoideum*. This cycle requires 24 hours under normal laboratory conditions.

(Cohen and Robertson, 1971). The net result of these signal and response properties is that the chemotactic signal is propagated outward and the cells move inward in pulses towards nearest secreting neighbors. Eventually the cells, which have become mutually adhesive (see below), are organized into coherent streams which converge to the autonomous signalling cells.

The biochemical elements of this system are beginning to be defined and understood (see review by Gerisch and Malchow, 1976). Malchow and Gerisch (1974) detected cAMP receptors, which increased at the cell surface as the amoebae entered into the aggregation-competent phase. Apparently cAMP is able to interact with these receptors without penetration into the cell (Malchow and Gerisch, 1973). A cell surface phosphodiesterase was also found to appear at the cell surface as the cells became aggregation-competent (Malchow *et al.,* 1972). This enzyme, which is clearly distinct from the cAMP receptor, is apparently responsible for enhancing the spatial gradient of cAMP seen by the cells and for terminating the signal following stimulation of the cell (Nanjundiah and Malchow, 1976). An extracellular phospho-diesterase which is present during the vegetative phase is partially or fully inactivated during aggregation (Gerisch *et al.*, 1972; Gerisch, 1976).

As described above, cAMP has a clear role as a chemotactic transmitter in the aggregation of *D. discoideum.* Early observations (Shaffer, 1957a; Bonner *et al.,*1969) also suggested that acrasin or a spatial gradient of acrasin induced cell cohesiveness. Recently, Gerisch *et al.* (1975) and Darmon *et al.* (1975) found that the pulses of cAMP in an aggregation field probably promote the differentiation of amoebae into an aggregation-competent state. Extracellular macromolecular substances may also play a role in this induction, perhaps in conjunction with cAMP (Gerisch and Huesgen, 1976; Alcantara and Bazill, 1976; Klein and Darmon, 1976).

During the later multicellular stages of *D. discoideum,* cAMP probably continues to function as an important extracellular control substance. Following up an early observation of Bonner (1949), Rubin and Robertson (1975) provided evidence that the tip of the slug continuously secretes acrasin, presumed to be cAMP. Through this secretion, the tip is thought to act as an embryonic organizer of the slug, defining its developmental axis and controlling the movement of posterior cells. As will be discussed below, high concentrations of cAMP in the anterior region of the slug may also control differentiation into stalk cells (Bonner, 1970).

(b) *Other species*

Most species exhibit streaming during aggregation. This spatial organization suggests signal relaying within the population, in which each cell can act as a local propagator of the chemotactic signal. Aggregation within other species is probably also initiated by autonomous signalling cells. In the case of *P. violaceum* (Shaffer, 1961) and *D. minutum* (Gerisch, 1965), these cells, called founder cells, are morphologically distinct from the cells that respond to them.

The nature of the chemotactic substances in other species is a subject of great current interest. *D. mucoroides* and *D. purpureum,* two close relatives of *D. discoideum,* appear to use cAMP as the transmitter during aggregation (Bonner *et al.,* 1972). However, amoebae of *P. pallidum, P. violaceum* and *D. minutum* were found to be insensitive to spatial gradients of cAMP. In *P. violaceum* amoebae were strongly attracted to their own aggregation centers but were only weakly attracted to a cAMP microelectrode (Jones and Robertson, 1976). Preliminary studies indicate that the chemotactic factor in this species might be a peptide (Wurster *et al.,* 1976).

Although apparently not important in aggregation, extracellular cAMP might play a regulating role in the later development of the *Polysphondylium* species. Late aggregation centers of *P. pallidum* and *P. violaceum* secreted cAMP (Bonner *et al.,* 1972) and the centers themselves were attracted to a cAMP-emitting electrode (Jones and Robertson, 1976). Furthermore, high concentrations of added cAMP disrupted the later development of these species (Jones and Robertson, 1976).

7.2.3 Adhesion

(a) *Temporal regulation*

Shaffer (1957b) made early studies of the development of cohesiveness with differentiation. He showed that vegetative and early pre-aggregation amoebae of several species of *Dictyostelium* and *Polysphondylium* were not cohesive; that is, cells coming into contact on a substratum showed little or no tendency to adhere to one another. After a variable period of starvation, depending on species and strain, the cells became organized into streams, which eventually moved to the center. At this stage, the cells were very cohesive, able to maintain their intercellular contacts despite various mechanical manipulations. The intercellular contacts were based on mechanical forces and were not the result of chemotactic clustering (Shaffer, 1957b).

Quantification of cohesiveness as cells differentiated from the vegetative phase to the aggregation-competent stage was described by Beug *et al.* (1973a). In their cohesion assay a suspension of dispersed cells was swirled passively forcing the cells into contact and allowing agglutinates to form. Agglutination was therefore dependent on the cohesiveness of the amoebae and not on their motility or chemotaxis. After equilibrium had been achieved, the degree of agglutination was measured by light scattering. Alternative measurement procedures quantified cohesiveness by determining the average agglutinate size or the percent reduction in single cells by a Coulter electronic particle counter (Rosen *et al.,* 1973, 1974) or by counting the proportion of aggregates of various sizes under the microscope (Gerisch, 1961; Alexander *et al.,* 1975).

Gerisch (1961) found, and Beug *et al.* (1973a) confirmed quantitatively, that EDTA had to be included in the cell suspension in order to eliminate a significant degree of agglomeration among vegetative amoebae. After 2 or 5 hours of starvation, (depending on the strain of *D. discoideum*), amoebae gained the ability to agglutinate in a gyrated suspension with EDTA present. Over the next 3 hours of development, this EDTA-resistant agglutination increased dramatically. It should be noted that cell ghosts exhibited the same agglutination characteristics as the cells from which they were derived (Sussman and Boschwitz, 1975). EDTA-resistant agglutination is thought to be a valid index of functionally-relevant cohesiveness, since its appearance in developing cells paralleled the capacity of the cells to enter into streams on a substratum (Beug *et al.,* 1973a). As will be discussed, quantitative cohesiveness assays have been an important tool in the investigation of cell adhesion mechanisms.

(b) *Morphology*

As amoebae move towards an aggregation center, they become aligned or polarized in their direction of motion. In *D. discoideum*, for example, the long axis of the cell is 5–6 times its width (Bonner, 1950). The elongated cells preferentially form end-to-end contacts giving rise to the single-file chains (Shaffer, 1964). An individual cell approaching a chain from the side will generally become incorporated into the chain rather than adhering laterally (Shaffer, 1964). These observations suggest that the anterior and posterior ends of elongated amoebae might be specialized adhesive sites. Lateral adhesions do occur, however, as chains merge into streams. These can be several cells wide and several layers deep. Cells entering into aggregation centers lose their elongated shapes and become isodiametric (Shaffer, 1975b). In the migrating slug (*D. discoideum*), the amoebae are either isodiametric or elongated at right angles to the longitudinal axis (Bonner, 1967).

Ultrastructural examination of aggregates of *D. discoideum* and *P. violaceum* revealed intercellular relationships resembling those found in unspecialized metazoan epithelium (Mercer and Shaffer, 1960). The cells were closely packed together with their trilaminar plasma membranes separated in many areas by a gap of approximately 200 Å. Small intercellular densities were observed which were suggestive of developing desmosomes. In the pseudoplasmodium of *D. discoideum*, Maeda and Takeuchi (1969) found that the anterior cells (prestalk) were more tightly packed than the posterior cells (prespore); as will be discussed, this observation correlated with the greater resistance of the former cells to disaggregation.

Freeze-fracture techniques have been used to examine plasma membrane intramembranous particles at various stages of *D. discoideum* development. Aldrich and Gregg (1973) found that at 8 hours of development, when the cells were cohesive, the membrane particles appeared to have coalesced to form larger units. It was suggested that the increased projection of the membrane particles might be important in cell adhesion. The addition of cAMP to vegetative amoebae led to a precocious appearance of the large particles (Gregg and Nesom, 1973). This observation might be related to the fact that added cAMP speeds up the development of *D. discoideum* under certain conditions.

(c) *Species-specific sorting-out*

Cellular slime molds of different species frequently occupy the same natural habitat. Yet spore caps and fruiting bodies found in the wild contain cells of only one species. This indicates that significant intermixture does not occur or at least does not persist. Raper and Thom (1941) provided the first systematic description of the segregation of different species in mixed cultures. When spores of any two of the species *D. discoideum, D. mucoroides* and *D. purpureum* were mixed, the fruiting bodies that eventually formed were typical of the species used with no evidence of intermixture. Following the actual process of sorting-out between *D. mucoroides* and *D. discoideum* was made possible by feeding the amoebae bacteria that contained a pigment which ones species could not degrade. Initially, amoebae of the two species converged to common aggregation centers. As aggregation advanced, the cells of the two species segregated within the centers and eventually gave rise to separate pseudoplasmodia, which culminated into separate characteristic fruiting bodies. The same basic sequence appeared to occur in the other two mixed pairings of the *Dictyostelium* sp. The initial co-aggregation of these pairs of species is not surprising, since as indicated above, they all appear to employ a primary chemotactic system based on cAMP. In contrast, when cells of *P. violaceum* were mixed with any of these *Dictyostelium* sp., there was no initial intermixture as the cells aggregated into completely distinct centers from the outset (Raper and Thom, 1941). Presumably, this lack of aggregation is due to the fact that *P. violaceum* cells, at least in their early stages of differentiation, do not secrete or respond to extracellular cAMP (Bonner *et al.*, 1972); but rather employ an entirely different chemotactic signal.

Bonner and Adams (1958), using a complex morphogenetic assay provided evidence that different slime molds sort out even after associations are forced by experimental manipulation. They removed the center of an aggregation pattern from the surrounding streams and replaced it with another center from the same or a different type of slime mold and then followed the fate of the two cell types. When the center and the streams were of the same strain, cells in the two segments became integrated and formed a normal single fruiting body. When the segments were of a different species or even of different strains of the same species, Bonner and Adams observed some degree of segregation of one of the following classes:

(1) two completely separate fruiting bodies;
(2) a double fruiting body with one type on top of the other; or
(3) a single fruiting body with spores segregated according to species within a single spore cap. Recently Garrod (personal communication) has shown sorting-out after mixed aggregates of different species or genera were produced in shaken suspension. In these experiments, the two cell populations were initially intermingled and then segregated within the aggregate.

The sorting-out of cells within morphogenetic aggregates or artificial agglutinates according to genera, species, or even strain is one of the most interesting properties of these organisms. Its mechanism is not presently known. One possibility is that the cells are segregated by chemotactic signalling within the multicellular mass. Another possibility is that aggregation is due to differential cell motility. Another mechanism for sorting-out supported by some experiments to be considered below is based on cell surface species-specific adhesion molecules.

(d) *Sorting-out and pattern formation*
In addition to species-specific sorting-out, there is evidence for selective cell association of two subpopulations of differentiated cells in *D. discoideum* as development progresses. The two subpopulations are the prestalk and prespore cells. It is clear that irrespective of the total number of cells constituting the slug, the anterior 20% will become stalk cells and the posterior 80% will become spore cells (Bonner, 1957). Furthermore, surgical removal of a part of the slug leads to re-establishment of this ratio of cell types (Raper, 1940). These results suggest that the determination of a cell as either a stalk or spore cell might simply depend on the relative axial position that the cell finds itself in the slug. However, considerable evidence indicates that the final positioning of cells is not random, but rather the cells sort-out to anterior

or posterior sites based on predispositions established during an earlier phase of the life cycle. Thus, in immunochemical studies, Takeuchi (1963) found that a sub-population of pre-aggregation amoebae, distributed at random in the field, stained with an antiserum raised against spores. With development of the aggregate, the amoebae that could be stained with the antiserum sorted-out to the posterior portion of the slug, whereas the unreactive cells became localized to the anterior section. Cell-sorting occurred during early stages of aggregation. At the time of slug formation, cell rearrangements were probably negligible (Farnsworth and Wolpert, 1971). Additional evidence in favor of 'tissue'-specific sorting-out (Garrod, 1974) has been provided by the experiments of Takeuchi (1969), Bonner *et al.* (1971), Leach *et al.* (1973) and Garrod and Forman (1977).

Given that sorting-out does occur, how can 'regulation' (Raper, 1940) be accounted for; that is, how is the proper stalk—spore ratio established in amputated segments of a slug? It is possible that a pattern-regulating mechanism exists which can respecify the fates of cells in a cut slug. Thus, cells would be predetermined as prestalk or prespore cells and sort-out to appropriate anterior-posterior positions. However, if the final position of a cell is not proper to its initial commitment (for example, after amputation) then it would be respecified (by the unknown patterning mechanism).

A major unsolved problem concerns the nature of the intrinsic property of the cells responsible for sorting-out. It has been suggested (Bonner, 1971; Garrod, 1974; Garrod and Forman, 1977) that differential adhesiveness of prestalk and prespore cell populations might underlie their sorting-out. The anterior—posterior segregation in the slug, rather than the classical sphere within a sphere configuration (Steinberg, 1964) is possibly due to the organizing influence of the slug tip. There are, in fact, indications of adhesive differences between anterior and posterior cells of the slug. Takeuchi and Yabuno (1970) found that anterior segments were much more difficult to dissociate into single cells than posterior regions. Morphologically, the cells in the anterior region were much more tightly packed (Maeda and Takeuchi, 1969). Furthermore, dissociated anterior cells were found to stick to a plastic substratum much more tenaciously than posterior cells (Yabuno, 1971). These observations should encourage further investigation of the relationship of adhesive properties to 'tissue' sorting-out.

(e) *Cell contact-dependent differentiation*

During the social stage of development in the cellular slime molds, cells within the multicellular assembly follow particular pathways of cytodifferentiation as a result of their communication and interaction with other cells and with their environment. We shall describe several experiments with the species *D. discoideum*, in which a beginning has been made in defining this complicated process and in providing possible explanations of the underlying mechanisms.

A central observation was that of Gregg (1971) who demonstrated that the developmental potential of isolated amoebae was drastically limited. Thus, vegetative cells or dissociated early-aggregate cells prevented from aggregation did not form organelles characteristic of late aggregation stages. If the blocked cells were allowed to aggregate, then the organelles appeared.

Studies of the effects of disaggregation and reaggregation on the expression of developmentally regulated enzymes have been particularly revealing. The enzyme UDPG pyrophosphorylase normally begins to accumulate in cells at about 12 hours of development, during the early pseudoplasmodium stage (Ashworth and Sussman, 1967). If pseudoplasmodia were mechanically dissociated and the cells plated at low density to prevent reaggregation, then the accumulation of this enzyme ceased (Newell *et al.*, 1971). However, if the cells were allowed to reaggregate — which they accomplished very quickly — they synthesized a constant amount of the enzyme. The amount synthesized was the same regardless of the amount made previously. Each successive disaggregation and reaggregation, up to three, resulted in the synthesis of an additional constant amount (or quantum) of the enzyme. This demonstration of the quantal control of enzyme synthesis by cellular interactions has been extended to three other functionally related enzymes (Newell *et al.*, 1972).

Although a number of other late enzymes do not display this elaborate quantal control in disaggregation-reaggregation experiments (Coston and Loomis, 1969; Loomis, 1969; and Firtel and Bonner, 1972), cellular interactions, of one kind or another, are required for the accumulation of several late enzyme. Grabel and Loomis (1977) showed that cells blocked from aggregation in sparse culture did not accumulate threonine deaminase, tyrosine transaminase, UDPG pyrophosphorylase, β-glucosidase or alkaline phosphatase.

Some progress was made in defining the nature of the cellular interactions that control the expression of these enzymes. Thus, cell communication within agglutinates formed in shaken suspension was sufficient for the expression of threonine deaminase and tyrosine transaminase but not for the three other enzymes. It appears that the regulation of these other enzymes depends on conditions found within the multicellular matrix of the pseudoplasmodium and that neither simple cell contact nor cell proximity in artificial agglutinates suffices (Grabel and Loomis, 1977).

What is the nature of the cell communication that guides or controls differentiation within the multicellular assembly? One possible mechanism might be based on diffusable inducers generated within the multicellular mass and passed between cells. This type of mechanism has gained support from work on stalk cell differentiation. Bonner (1970) showed that exogenous cAMP induced isolated amoebae to become stalk cells. This induction may also involve a diffusable substance, released by cells, that acts in conjunction with the cyclic nucleotide (Town *et al.*, 1976). In a number of studies (Bonner, 1949; Pan *et al.*, 1974; Brenner, 1977), cAMP was shown to be concentrated in the anterior of the pseudoplasmodium which is destined to become stalk in the mature fruiting body. These results suggest that cAMP may be an inducer of stalk cell formation during the pseudoplasmodial stage.

Another mechanism for cell communication might involve cell contact-induced cytodifferentiation. In this case, actual contact would trigger programs within the cells without the passage of substances. Evidence for this comes from the demonstration that exogenously supplied membranes from aggregation-phase amoebae affected the expression of several developmentally regulated enzymes within cells while blocking the aggregation and subsequent morphogenesis of the cells (Smart and Tuchman, 1976; Tuchman *et al.,* 1976). The presumption is that the membranes were acting by binding in a specific manner to the surfaces of cells rather than by release of factors that associated with or entered the cells. Of particular interest is the finding that membrane-treated cells accumulated alkaline phosphatase nine hours precociously and to twice the normal level (Tuchman *et al.,* 1976). Elucidation of the mechanism of cell adhesion should permit further study of cytodifferentiation regulated by cell contact.

7.3 MECHANISMS OF ADHESION

7.3.1 Advantages of studying cellular slime molds

The cellular slime molds are a favorable system for studying the problem of cell adhesion. The main reason is that there is a clear developmental transition, readily controlled by the availability of food, between a non-social, non-cohesive state and a social, cohesive one. Over the several-hour period in which cells differentiate to an aggregation-competent state, one can study accompanying cell surface changes by morphological, immunological, or biochemical techniques. These studies are aided by the fact that slime molds are easily and inexpensively cultured, and large quantities of cells at any stage of development can be obtained. Furthermore, analysis of adhesion mechanisms may be assisted by aggregateless mutants, which are readily generated. Several of those already studied appear to have general developmental blocks at stages prior to aggregation (see Loomis, 1975). Undoubtedly, others will be identified having discrete and specific lesions in the cell adhesion apparatus.

Despite the advantages of cellular slime molds, an understanding of the cellular adhesion they display is presently limited. In the ensuing sections we will review a variety of studies, all of which are relevant to the molecular basis of cellular adhesion in these organisms.

7.3.2 Cell surface chemistry in development of cell cohesiveness

(a) *Biophysical studies*
Biophysical studies of plasma membrane fluidity or electrostatic properties with slime mold development show no definitive changes that can be related to mechanisms of aggregation on adhesion.

Von Dreele and Williams (1977) measured plasma membrane fluidity at various stages of development in *Dictyostelium discoideum.* Despite shifts in the polar and neutral lipid composition (Ellingson, 1974; Long and Coe, 1974), the fluidity of the bulk lipids remained constant at a relatively high level. Membrane fluidity would therefore not appear to be a critical determining factor in aggregation or cell cohesion.

The lyophobic colloid stability theory of cell adhesion (see Curtis, 1967) states that stable cell contact results from a balance between electrostatic repulsive forces and attractive London—Van der Waals forces. In slime molds, change in one or both of these forces might account for the regulation of cohesiveness during development. In *D. discoideum*, the electrophoretic mobility of cells (reflecting electrostatic surface potential) was found to decrease between the non-social and social phases (Garrod and Gingell, 1970; Yabuno, 1970; Lee, 1972a). However, the decrease was not specifically correlated with aggregation, as there was not an abrupt change at the onset of aggregation competence. Furthermore, cells in certain defined media (glucose or amino acids plus vitamins) were able to form aggregates, albeit somewhat loose, despite the fact that their electrophoretic mobility had not decreased (Lee, 1972b). From these experiments, Lee (1972b) suggested that a reduced surface charge might contribute to the strength of cohesion, but it is not the critical factor regulating stage-specific cohesion. He speculated that discrete intercellular binding substances, undetected by cell electrophoresis, are the determining factors.

(b) *Discrete cell surface protein changes with development of cohesiveness*
Early evidence that cell surface proteins might play a role in cell cohesion was provided by Takeuchi and Yabuno (1970). They demonstrated that pronase as well as several other proteolytic enzymes, working in conjunction with a disulfide reducing agent, were effective in dissociating the pseudoplasmodium of *D. discoideum* into single viable amoebae. In addition, trypsin-treatment of mechanically dissociated slug cells eliminated their cohesiveness, as measured by agglutination in the presence of EDTA (Alexander *et al.,* 1975). Although protease effects on living cells are complex, these experiments suggested that cell surface proteins might participate in cell contact.

Adopting the working hypothesis that membrane proteins are involved, several investigators have attempted to define the relevant proteins by identifying those proteins that change during the acquisition of aggregation competence. One technique used was lactoperoxidase-mediated radio-iodination of intact cells, which labels exposed tyrosine residues on external membrane proteins. In applying this method to *D. discoideum*, Smart and Hynes (1974) identified several labeled proteins on SDS-polyacrylamide gels. Four of these bands either increased or appeared while one band decreased over the first 12 hours of development. Siu *et al.* (1975) also found a changing pattern of labeled proteins by this technique, but there was poor correspondence between the two studies in the molecular weights of the bands and in their developmental changes. Both groups did concur, however, that a polypeptide with a molecular weight of 130 000 became accessible to labeling by 12 hours of differentiation. According to Smart and Hynes (1974), this labeled component was present on cells in the organized pseudoplasmodium but was not on aggregation-competent cells differentiated in suspension culture. Since it was absent on these cohesive cells and its developmental onset was late, this protein, while very interesting, is probably not involved in primary cell adhesion.

Another method for detecting protein changes has been to determine the polypeptide composition of isolated plasma membranes from the different stages of *D. discoideum*. Using high-resolution gradient polyacrylamide SDS gels, Hoffman and McMahon (1977) resolved over 55 Coomassie Blue staining bands in plasma membranes from vegetative cells. Some of these bands might have been adsorbed cytoplasmic proteins and not true membrane proteins. During the first 12 hours of development, 8 bands diminished, and 5 bands either increased or appeared for the first time. Three of these 5 bands appeared to be external membrane proteins, since pronase treatment of intact 12-hour cells destroyed them. There was no apparent-relationship between these three and the external proteins labeled in the lactoperoxidase experiments of Smart and Hynes (1974). A few of the polypeptides that were present in both growth-phase and aggregation-phase membranes were more pronase-sensitive in the differentiated cells, suggesting the possibility of increased exposure of these proteins in the aggregation-competent cells. Conceivably, topographical change in a cell surface protein might be the basis for its functional activation.

This type of analysis has identified several interesting proteins of two classes: those that increased in amount as cells became cohesive and those that became more pronase-sensitive in intact differentiated cells. The relationship of these changes to cell cohesion or other aspects of cell surface differentiation remains to be determined.

(c) *Discrete glycoprotein changes*

There is a widespread belief that oligosaccharides, which are abundant in cell surface glycoproteins and glycolipids, may play a specific role in intercellular adhesion. Changes in cell surface glycoproteins of slime molds occur concurrently with the asquisition of aggregation competence. Hoffman and McMahon (1977) analyzed the glycoprotein components of *D. discoideum* plasma membranes by Periodic Acid Schiff staining of SDS-polyacrylamide-gradient gels. Vegetative plasma membranes contained over 15 glycoproteins and by 12 hours of development, one of these bands had disappeared, two had diminished, and three new bands had appeared. One surprising result was that the glycoproteins in both stages were largely insensitive to pronase treatment of intact cells. It is not known whether the glycoprotein changes that were detected by this method are important in cell adhesion.

The changing cell surface of *D. discoideum* has also been studied with the use of plant lectins, which are specific carbohydrate-binding proteins that agglutinate cells. Weeks (1973) and Kawai and Takeuchi (1976) found that concanavalin A (which binds predominantly to α-glucopyranosyl or α-mannopyranosyl residues) agglutinated vegetative cells to a greater degree than aggregation-competent cells. The susceptibility to Con A agglutination decreased gradually over the first 10 hours of development (Weeks and Weeks, 1975). Using glutaraldehyde-fixed cells in order to eliminate background agglutination, Reitherman *et al.* (1975) found that both Wheat Germ

Agglutinin (which binds predominantly with *N*-acetylglucosamine residues) and Con A agglutinated growth-phase cells more effectively than differentiated cells. On the other hand the *Ricinus communis* Agglutinin I (D-galactopyranosyl residues) agglutinated differentiated cells more effectively than vegetative cells.

Interpretation of these studies is difficult, since agglutination is a complex phenomenon dependent on many diverse factors including nature and number of binding sites, affinity of binding, arrangement and mobility of sites, general cell surface topography, and electrostatic surface charge (Nicolson, 1974). One can safely attribute changes in lectin-mediated agglutination to changes in some component of the plasma membrane or associated structures; but the exact molecular basis of the change cannot be discerned without additional information.

Direct study of cell surface 'receptors' for lectins can be made with binding studies. Labeled Con A binding to slime mold cells has been studied by several investigators. They found a small increase or no change at all (Weeks, 1975; Geltowsky *et al.*, 1976; Darmon and Klein, 1976) in the number of cell surface binding sites in vegetative and aggregation-competent cells. On the basis of sites per unit surface area, there was a marked increase with differentiation (Weeks, 1975; Geltowsky *et al.*, 1976). Decreasing agglutination with differentiation was therefore not due to either a diminished number or a diminished density of Con A receptors. A Scatchard analysis (Weeks, 1975) suggested that the critical factor might be changing binding affinities for the lectin; as cells differentiated, a majority of the receptors decreased substantially in their affinity for Con A. Presumably, the different affinities reflected a changing cell surface carbohydrate composition, as a result of either synthesis of new oligosaccharides or alteration of pre-existent chains.

Another approach to studying changing Con A receptors in developing *D. discoideum* was taken by Geltowsky *et al.* (1976). They purified detergent-solubilized receptors, labeled at the cell surface by the lactoperoxidase method, on a column containing covalently bound Con A. As analyzed by SDS polyacrylamide gel electrophoresis, there were 15 distinct receptors found in vegetative cells. Between 6 and 18 hours of development, one band of molecular weight 150 000 increased dramatically while another band (180 000) decreased. The 150 000 dalton band might correspond to band 2, identified in the experiments of Smart and Hynes (1974).

In yet another approach, West and McMahon (1977) used the 'lectin-electrophoresis' method, in which Con A was substituted for antibody in crossed-immunoelectrophoresis. By this method, they established that the majority of glycoproteins in vegetative plasma membranes revealed by PAS-staining of SDS gels, were Con A receptors. These investigators employed an even higher resolution technique, in which fluorescent Con A was allowed to bind to individual components in a SDS gel. Over 35 Con A-binding glycoproteins ranging in molecular weight from 8000 to 313 000, were found in vegetative plasma membranes. Most of these receptors were also present in aggregation-phase membranes (12-hour cells), but several new receptors had appeared and a few had disappeared. There were no notable changes at positions in the gel corresponding to 150 000 and 180 000 daltons. It is possibly of significance that Geltowsky *et al.* employed the NC–4 strain while West and McMahon used the axenic mutant A3. The binding of Con A to such a large number of membrane glycoproteins may reflect the fact that this lectin, in addition to recognizing certain terminal residues, also binds to internal mannose residues, which are commonly found in the core structure of many glycoproteins (see Liener, 1976).

The plant lectin studies have revealed cell surface changes in differentiating *D. discoideum.* In the case of Con A, at least, there were significant changes in discrete cell surface glycoproteins. A key question is whether lectin receptors, in particular those that are developmentally regulated, function in cell adhesion. One possible way to investigate their functional role would be to determine the effects of blocking these receptors with added lectin on the developmental program. In such an experiment Con A was found to delay significantly the onset of aggregation in *D. discoideum* (Gillette and Filosa, 1973; Weeks and Weeks, 1975; Darmon and Klein, 1976). The delay was probably not due to a simple masking of receptors, since succinyl-Con A, a lectin derivative of reduced valency but with the same binding specificity, did not affect development (Weeks and Weeks, 1975; Darmon and Klein, 1976). The inhibitory activity of intact Con A was probably related to its dramatic general effects on cell surface architecture of living cells. Gillette *et al.* (1974) demonstrated that bound Con A redistributed from a uniform layer to form a cap-like structure on the amoeba. In investigations with the scanning electron microscope, Molday *et al.* (1976) showed that Con A caused clustering and capping of its receptors as well as a dramatic rounding-up of the cell and clustering of microvilli. Ultimately, the Con A was probably endocytosed (Darman and Klein, 1976; Grabel and Farnsworth, 1977). In view of the drastic surface remodeling induced by Con A, it is not surprising that aggregation was delayed. Succinyl-Con A, because of its reduced valency, would not be expected to have these effects. This is presumably the reason it had no effect on development.

7.3.3 Immunological studies

(a) *Development – specific antigens*
One strategy used for studying cell cohesion is to obtain antisera that are specific to cohesive cells, then attempt to use these antisera to determine cell surface antigens specifically involved in cohesion. Gregg (1956) provided early evidence that new antigens appeared on the surface of *D. discoideum, D. purpureum* and *P. violaceum* amoebae between the vegetative and aggregation phases. Other studies (Gregg and Trygstad, 1958) indicated that several aggregateless mutants of *D. discoideum* had altered surface antigens relative to the wild type.

Sonneborn *et al.* (1964) made the first attempt to assess the role of new surface antigens in the process of aggregation. They employed an aggregation-specific antibody: raised against *D. discoideum* aggregates and then exhaustively absorbed with vegetative cells. By complement fixation, a particulate-bound antigen was detected which increased 100-fold by the onset of aggregation. Moreover, the antiserum blocked aggregation of cells on an agar surface without impairing their viability. However, studies with a mutant of *D. discoideum* made it very unlikely that the antibody was directly affecting adhesion between cells. This mutant aggregates at high density as a result of random encounters between cells, rather than oriented chemotaxis. The clumps that form culminate in fruiting bodies. The aggregation-specific antigen was not detected in this mutant. Furthermore, the antiserum did not block its aggregation or later development. It would appear, therefore, that the developmentally regulated antigen identified by Sonneborn *et al.* (1964) is probably important in chemotaxis or cell motility, but not in cell contact formation. In subsequent work, Goidl *et al.* (1972) found that antibody against phosphodiesterase blocked the aggregation of *D. discoideum,* presumably by interfering with some aspect of chemotaxis. Clearly, antibodies that block aggregation of slime mold cells need not be directed against molecules that directly mediate cell cohesion.

(b) *Adhesion-blocking antibody*

The most comprehensive approach to identifying the relevant antigens in cell adhesion has been taken by Gerisch, Beug and their colleagues. High titer antibodies were raised against crude extracts (total homogenate or particulate fraction) of aggregation-competent *D. discoideum* cells (Beug *et al.,* 1970). The antibodies were degraded to univalent Fab fragments in order to avoid immune agglutination of cells or a redistribution of surface antigens into clusters or caps. High concentrations of specific Fab (greater than 2 mg ml^{-1}) blocked the aggregation of differentiated amoebae on a substratum (Beug *et al.,* 1970). This inhibition was achieved without noticeable effect on the shape, pseudopodial activity, motility or chemotactic responsiveness of the cells. Further evidence that this Fab blocked cell adhesion rather than some other process was the fact that it inhibited the passive agglutination of aggregation-competent cells in a gyrated suspension (Beug *et al.,* 1973a). As discussed before, such agglutination depends on the ability of cells to form stable cell contacts and not on chemotaxis or cell motility. Not only did this Fab block the formation of new cell contacts, whether the cells were on a substratum or in a swirled suspension, but it also dissociated pre-existing cell contacts in an organized pseudoplasmodium (Beug *et al.,* 1971). In the presence of a high concentration of Fab (32 mg ml^{-1}) after several hours of incubation, sliced-up slugs were dissociated into single, viable cells.

(c) *'Contact Sites' A and B*

Beug *et al.* (1973a) have distinguished two independent systems of cell adhesion in *D. discoideum* by several means. Early work (Gerisch, 1961) discussed before, showed that both vegetative and aggregation-competent amoebae agglutinated in gyrated suspension. However, the agglutination of vegetative cells was inhibited almost completely by EDTA, whereas the agglutination of the differentiated cells persisted in the presence of this agent. As discussed above, the appearance of EDTA-resistant agglutination in developing cells was correlated with the ability of cells to form firm end-to-end contacts on a substratum. Beug *et al.* (1973a) prepared univalent antibodies from pooled antisera obtained from many rabbits repeatedly immunized with broken differentiated cells and exhaustively absorbed with broken vegetative cells (aggregation-specific Fab). This reagent blocked the EDTA-resistant agglutination of aggregation-competent cells but did not affect the EDTA-sensitive agglutination of vegetative cells. The target antigens to which the agglutination-inhibiting Fab bound were operationally defined as 'Contact Sites' A (CS A). Fab prepared against vegetative cells (and absorbed against a small quantity of differentiated cells to sharpen its specificity) blocked EDTA-sensitive agglutination but had no effect on the EDTA-resistant agglutination. The target antigens in this case were defined as 'Contact Sites' B (CS B). Because of the crude nature of the immunizing antigens, only a fraction of the antibodies in the antisera would be reactive with the actual 'contact site' antigens.

It was possible to quantify 'Contact Sites' A and B in cells at various stages by determining loss of agglutination-inhibiting activity in appropriate Fab fractions after absorption with particulate fractions derived from the cells. Before 4 hours of development, CS A were undetectable. Between 4 and 9 hours, there was a steep increase in the amount of CS A, which closely paralleled the development of aggregation competence, as measured by EDTA-resistant agglutination. Intact aggregation-competent amoebae could absorb out CS A-specific Fab, indicating that the relevant antigens were available at the cell surface (Beug *et al.*, 1973b). In contrast to CS A, CS B did not vary substantially in amount over the first 9 hours of development.

Absorption experiments with aggregation-competent *P. pallidum* cells indicated that CS A were species-specific (Beug *et al.*, 1973a).

(d) *Topography and discreteness of 'contact sites'*

CS A and CS B-specific Fab have dramatically different effects on the spatial pattern of cell aggregation on a substratum (Beug *et al.*, 1973a). Antagonism of A antigens resulted in a loose lateral association of the elongated amoebae. Inhibition of B sites by either specific Fab or EDTA produced end-to-end assemblies of cells in either chains or rosette patterns. Possible localization of CS A antigens to cell ends, which might be inferred from these results, could not be established by immunocyto-chemical studies. Aggregation-specific antibody bound evenly all around the cell (Beug *et al.*, 1973b; Gerisch *et al.*, 1974). Possibly, antibody species directed to surface antigens other than the actual CS A sites might have obscured the A site localization.

A very interesting feature of the cell surface antigens recognized by the aggregation-specific antibody was that their average extension from the plasma membrane was 30–40 Å beyond glycolipid antigens recognized by another antibody (Gerisch *et al.*, 1974). This result was based on measurements of the distance between the outer plasma membrane dense layer and ferritin in the second antibody layer. The distribution of distances was very broad and it could not be determined whether CS A, the candidates for intercellular adhesion factors, were among the most extended antigens.

The question of the discreteness of the two classes of contact sites has been explored. Beug *et al.* (1973b) found that an aggregation-competent cell bound no more than 3×10^5 aggregation-specific Fab molecules. Because of the polyspecificity of the antibody, this number would represent an upper limit estimate of the actual number of CS A. The binding of a maximal amount of this Fab would have covered no more than 2% of the cell surface (Beug *et al.*, 1973b). On the other hand, 2.5×10^6 Fab molecules of 'anticarbohydrate' specificity (directed against cell surface antigens I and II) could bind to each aggregation-competent cell without affecting EDTA-resistant agglutination (CS A). Similarly, CS B were completely inhibited by binding of 2×10^6 molecules per cell of specific Fab, but binding of 'anticarbohydrate' Fab to the same number of sites had only a weak effect. Thus, in both cases, inhibition of intercellular adhesion was achieved by the interaction of Fab molecules with a discrete subpopulation of cell surface antigens.

(e) *Role of 'contact sites' in adhesion*

Aggregateless mutants have been used in an attempt to evaluate the significance of the CS A system. In one class of mutants, neither CS A nor the ability to form EDTA-resistant agglutinates developed with starvation (Beug *et al.*, 1973a, Gerisch *et al.*, 1974). Most members of this group probably suffer from a block in differentiation at some early stage in development. A mutant with a lesion in the structural gene for a CS A component would, of course, be much more valuable than pleiotropic control variants. Another group of aggregation-defective mutants exhibited no loss of CS A absorbing activity, indicating that either CS A were serologically active but functionally inactive, or that the mutant lacked a membrane component required for adhesion, for which there was inadequate inhibiting Fab species. Finally, a third class of mutants was found in which the cells were capable of absorbing only a part of CS A blocking Fab (Gerisch *et al.*, 1974). The residual activity in this absorbed Fab which could completely block EDTA-insensitive agglutination was removed by absorption with wild-type cells. These results suggested that there are at least two components of CS A, each antagonized by a specific Fab. Both components are required for adhesion, and interference with either is sufficient to block adhesion. Gerisch *et al.* (1974) speculated that these components might represent complementary sites on adjacent cells, whose interaction mediates cell-to-cell contact.

This model is both attractive and plausible. However, as appreciated by Beug *et al.* (1973a), the available evidence does not allow the conclusion that these immunologically-defined 'contact sites' directly participate in cell adhesion. For example, it is possible that 'contact sites' are regulatory factors controlling the activity or display of true adhesion sites.

(f) *Chemical nature of 'contact sites'*

Further information on the role of 'contact sites' in adhesion awaits purification and characterization of the antigens. In preliminary experiments (Huesgen and Gerisch, 1975), CS A were solubilized from crude membrane fractions by sodium deoxycholate and purified 160-fold by chromatography on DEAE-cellulose and G-200 Sephadex. Activity was followed by measuring neutralization of agglutination-inhibiting Fab. The physical fractionation of the solubilized membranes clearly separated CS A from cell surface phosphodiesterase, which has a similar temporal regulation. Based on molecular sieving, the molecular weight of CS A was estimated to be 130 000 with no indication of size heterogeneity. The activity of solubilized CS A was sensitive to both periodate and pronase treatments, suggesting that serological activity depended on protein and carbohydrate (Gerisch, 1975). As yet, the isolation of purified homogeneous CS A has not been reported.

7.3.4 Endogenous cell surface lectins

(a) *Developmentally regulated hemagglutinin in D. discoideum*

The work (Rosen, 1972) began with an anomalous result in a passive hemagglutination assay. Normally, in this procedure, protein antigens are non-specifically absorbed to erythrocytes (fresh or fixed) pretreated with tannic acid. Specific antibodies are then detected and measured by their hemagglutination activity against the coated cells. When tannic-acid treated, formalin-fixed sheep erythrocytes were exposed to a crude soluble extract of *D. discoideum* slugs, the erythrocytes spontaneously agglutinated without the addition of antiserum. The key observation was that the level of agglutination activity in the slug extracts (per milligram protein) was two orders of magnitude higher than in extracts of vegetative amoebae. Furthermore, over the first 12 hours of development, specific hemagglutination activity in cell extracts increased over 400-fold, with a precipitous rise between 6 and 9 hours after food deprivation. The substance responsible for the agglutination activity was shown to be heat-labile, trypsin-sensitive, and non-dialysable, indicating that it contained protein. The striking developmental regulation of this agglutinin suggested its possible importance in cell adhesion.

(b) *Identification of agglutinins as lectins. Parallel appearance of lectins and cohesiveness*

Subsequent investigation (Rosen *et al.*, 1973) established that it was not necessary to treat erythrocytes with tannic acid or formaldehyde in order to detect agglutination activity. However, formalinized erythrocytes were generally employed in assays, since these cells were as sensitive as fresh erythrocytes and were much more convenient.

The nature of the hemagglutinin was revealed by the finding that agglutination activity produced by slime mold extracts could be blocked by specific simple saccharides. For example, the agglutination of formalinized sheep erythrocytes by a crude *D. discoideum* extract was inhibited by *N*-acetyl-D-galactosamine (Rosen *et al.*, 1973). Other saccharides such as D-galactose also inhibited agglutination but were less effective. Several sugars including D-glucose and *N*-acetyl-D-glucosamine exhibited no inhibitory activity. Appropriate simple sugars also blocked agglutination produced by extracts of *P. pallidum*.

These results and others discussed below indicated that the slime mold hemagglutinins were members of the class of carbohydrate-binding proteins known as lectins. These substances are a highly diverse family of proteins, isolated from a wide variety of plant, microbe, and animal sources, which have the ability to agglutinate erythrocytes and other cell types (Sharon and Lis, 1972). A lectin agglutinates cells by attaching via two or more of its carbohydrate-binding sites to oligosaccharides exposed on adjacent cells. Specific simple sugars compete for the binding sites and thereby inhibit agglutination.

The appearance of lectin-mediated agglutination activity in soluble extracts of differentiating *D. discoideum* (strain NC—4, bacterial — grown) or *P. pallidum* cells paralleled the development of cellular cohesiveness (Rosen *et al.*, 1973, 1974). Appearance of both cohesiveness and lectin could be blocked by cycloheximide, a protein synthesis inhibitor. Cohesiveness was quantified by measuring EDTA-resistant agglutination of cells in a gyrated suspension by a method similar to that of Beug *et al.* (1973a). In *D. discoideum* both lectin activity and cell cohesiveness, starting from non-measurable levels, increased markedly between 6 and 9 hours of development. In *P. pallidum*, in contrast to *D. discoideum*, growth-phase cells generally had a significant amount of lectin activity. After 4 hours of development, this level increased about 10-fold in parallel with the acquisition of cell cohesiveness.

(c) *Purification and characterization of lectins*
Because of its affinity for galactosyl residues, the agglutinin from *D. discoideum* could be purified by affinity chromatography on Sepharose 4B, a linear polymer of galactose residues (Simpson *et al.*, 1974). The purified agglutinin, eluted from the matrix by free galactose, was resolved by DEAE-cellulose into separate species called discoidin I and discoidin II (Frazier *et al.*, 1975a). The proteins were found to be tetramers with molecular weights of about 100 000 (Simpson *et al.*, 1974; Reitherman *et al.*, 1975), each made up of 4 identical subunits of 26 000 (discoidin I) or 24 000 (discoidin II) daltons (Frazier *et al.*, 1975a). Discoidin I and discoidin II appeared to be separate gene products, since they had clearly distinct amino acid compositions and tryptic peptide maps. The carbohydrate-binding specificities of the two proteins were somewhat different (Frazier *et al.*, 1975a), as assessed by the relative ability of simple sugars to inhibit hemagglutination by the isolated lectins. The lectins exhibited different relative activities in agglutinating sheep, rabbit and human erythrocytes, presumably reflecting differences in their carbohydrate-binding sites.

Discoidins I and II also differed in their developmental regulation, as reflected in the amounts extracted from cells at different stages (Frazier *et al.*, 1975a). Both lectins were undetectable in extracts of vegetative cells. However, between 6 and 12 hours, discoidin I increased 20-fold whereas discoidin II increased only 3-fold to a level that was only 10% of that reached by discoidin I.

The agglutinin from *P. pallidum* was first purified by adsorption to fixed erythrocytes and elution with a specific saccharide (Simpson *et al.*, 1975). The purified lectin, called pallidin, had a different isoelectric point, amino acid composition and peptide map than discoidin I or II (Simpson *et al.*, 1975; Frazier *et al.*, 1975b). The weight average molecular weight was 250 000, with a subunit size of 25 000. Sugar hapten inhibition studies demonstrated that the carbohydrate-binding specificity of pallidin was clearly distinct from that of discoidin I or II (Rosen *et al.*, 1974; Frazier *et al.*, 1975a). For example, *N*-acetyl-D-galactosamine was an excellent inhibitor of discoidin I and II but a poor inhibitor of pallidin. Lactose was the best simple saccharide inhibitor of pallidin. As with the discoidins, there was no detectable hexosamine or neutral sugar associated with pallidin (Simpson *et al.*, 1974, 1975).

The use of acid-treated Sepharose rather than fixed erythrocytes as the affinity absorbent has allowed the identification of 4 distinct isoelectric variants of pallidin (Rosen and Kaur, unpublished observations). The physicochemical characteristics of these different forms are currently under investigation.

The lectins from *D. discoideum* and *P. pallidum* are major cell proteins, comprising 2—3% of the total cellular protein of aggregation-competent amoebae. Each cohesive cell contains 2×10^6 pallidin molecules in the case of *P. pallidum* and 6×10^6 discoidin molecules (I and II) in the case of *D. discoideum*. These estimates are based on extractable lectin and must therefore be considered lower limits.

Four other species of slime molds have been shown to contain lectins (Rosen *et al.*, 1975). In crude extracts of aggregation-competent cells of six species including *D. discoideum* and *P. pallidum*, the lectin activities were distinguishable by sugar hapten inhibition of hemagglutination. However, since these lectin activities might have reflected composite activities of several subspecies, one cannot conclude that each slime mold possesses lectins with unique binding specificities. Determining the relationship of the slime mold lectins will require their isolation and characterization. Work has begun in this direction using Sepharose (Frazier *et al.*, 1975b) or fixed erythrocytes as the affinity adsorbants for purification. The major lectins from *D. mucoroides* (mucoroidin) and *D. purpureum* (purpurin) show many similarities to the discoidins but they are each distinguishable physicochemically, immunologically and by the relative potency of a series of sugars to inhibit their hemagglutination activity (Barondes and Haywood, unpublished). By this last criterion, discoidin I and mucoroidin are much more like each other than like purpurin.

(d) *Cell surface location of lectins*

Studies from several laboratories, employing a variety of techniques, have established that the lectins of *D. discoideum* and *P. pallidum* are present on the cell surface of cohesive amoebae. This was first indicated by the finding that intact cohesive amoebae formed rosettes with erythrocytes, which could be blocked by lectin-specific sugars but not by non-specific sugars (Rosen *et al.*, 1973; 1974; Chang *et al.*, 1977). In *D. discoideum*, the ability to form rosettes was clearly developmentally regulated, since erythrocytes exhibited little binding to vegetative cells but substantially more binding to cohesive amoebae. Rosette formation was presumed to be due to cell surface lectin on slime molds recognizing and attaching to oligosaccharides on erythrocytes.

Using antibodies to purified discoidin, Chang *et al.* (1975) demonstrated its presence on the surface of cohesive cells by both immunofluorescent and immuno-ferritin labeling techniques. Vegetative amoebae, briefly fixed and treated with rabbit antidiscoidin followed by fluorescent-labeled goat antirabbit antibody, showed no specific staining, whereas 12-hour differentiated cells exhibited substantial fluorescence. The staining was diffusely distributed over the entire surface of the cells with no suggestion of preferential localization in regions of intercellular contact. Staining became intense 8–10 hours after the initiation of differentiation (Strain NC–4, bacterial-grown). When living 9-hour differentiated cells were reacted with antibodies, labeling was initially diffuse over the surface. However, with incubation at room temperature, label collected into patches, which eventually consolidated into one large cap per cell. This antibody-induced redistribution indicated that the lectin (or a molecule in the cell membrane to which it is bound) could move within the plane of the membrane. Studies with ferritin-conjugated antibody confirmed the general results of the fluorescent antibody studies. Labeling was seen at the ends and sides of aggregation-competent cells. As the antibodies were prepared to a mixture of discoidin I and II, these immunocytochemical studies did not distinguish the two lectins.

Similar studies with antibodies to pallidin (Chang *et al.*, 1977) demonstrated the presence of this lectin on cohesive *P. pallidum* cells. In contrast to *D. discoideum*, detectable levels of lectin were found on vegetative cells. These amoebae also exhibited a degree of EDTA-resistant agglutination in gyrated suspension. As the cells differentiated, both surface lectin and cohesiveness increased.

The lectin of differentiated cells was not restricted to the cell surface (Chang *et al.*, 1977). When *P. pallidum* cells were made permeable to antibody, considerable staining with antipallidin was found associated with regions of rough endoplasmic reticulum. The relative amounts inside the cell and on the surface are not known. Furthermore, the route of the lectin to the cell surface and the mechanism of its incorporation into or onto the plasma membrane remain to be determined.

Radio-iodination of intact cells in combination with specific immunoprecipitation by antidiscoidin has confirmed both the surface location of discoidin and its developmental regulation at this site (Siu *et al.*, 1976). Cells at different stages of development were radio-iodinated by the lactoperoxidase technique; detergent extracts of the labeled cells were treated with antibody, and immune precipitates were analyzed on SDS gels to ascertain the amount of iodinated discoidin. In accord with previous immunocytochemical results, discoidin was absent on bacterial-grown vegetative cells (strains A3 and NC–4), appeared by 5 hours of development and increased through 9 hours. Lectin was found on the cell surface through the culmination stage. Expression of cell surface lectin required both RNA and protein synthesis. Analysis of mutants by this technique revealed that 2 of 3 non-cohesive lines did not express lectin on the cell surface, while the third had reduced levels relative to the wild type.

Rosen *et al.* (1973) found large amounts of discoidin in extracts of axenically grown vegetative cells (Strain A3). These cells in contrast to bacterial – grown cells also exhibited EDTA-insensitive agglutination. Siu *et al.* (1976) were able to detect considerable lectin on the surface of these cells during their exponential growth phase. As suggested by these workers, cells grown in liquid media might be partially starved, and certain of the early developmental programs might begin precociously.

Discoidin has also been detected in preparations of purified plasma membranes by SDS gel analysis (Hoffman and McMahon, 1977). Purified membranes from growth phase amoebae (axenically grown A3) and 12-hour differentiated cells both contained a protein that co-migrated with authentic discoidin on SDS gels. The only notable change was that the pronase sensitivity of the protein in intact cells appeared to increase during development, suggesting the possibility of increased exposure on the surface of the cells. The finding that these co-migrating bands in vegetative and 12-hour membranes could be eluted from the gels and renatured to active galactose-binding proteins established that they contained discoidin (West and McMahon, 1977).

An important problem for future work concerns the nature of the association between cell surface lectin and the plasma membrane. It is not known whether the lectin is a peripheral membrane protein attached to the cell of origin by its carbohydrate-binding function or an integral membrane protein intercalated into the lipid bilayer.

(e) *Cell surface receptors for lectins*

If slime mold lectins mediate cell cohesion, it is necessary not only that they be present at the cell surface but also that the cell surface contain appropriate receptors for these carbohydrate-binding proteins. Agglutination studies provided the first evidence that such receptors existed. Aggregation-competent *P. pallidum* amoebae, heat-treated to reduce endogenous cohesiveness, were agglutinated by added pallidin in a gyrated suspension (Rosen *et al.*, 1974). D-galactose, a specific inhibitor of the lectin blocked the agglutination, but *D*-glucose, a non-specific sugar, had no effect.

Glutaraldehyde-fixed *D. discoideum* cells of appropriate developmental stage were agglutinated by discoidin I or II (Reitherman *et al.*, 1975). Fixation was employed to block endogenous cohesiveness without altering oligosaccharide chains, potential receptors for the lectins. The agglutinability of fixed cells by added discoidin I or II increased dramatically between 0 and 9 hours of development (bacterial-grown, NC–4). Whereas 0-hour cells were not agglutinated and 3-hour cells were only slightly agglutinated, the lectins agglutinated 6-hour cells markedly and 9-hour cells to an even greater extent. Pallidin and *Ricinus communis* I, which like the discoidins, bind D-galactopyranosyl residues, also agglutinated 9-hour cells, but not vegetative cells. This increasing agglutinability was not, however, observed for all lectins. Wheat Germ Agglutinin and Concanavalin A, plant lectins with binding specificities dissimilar to the discoidins, agglutinated vegetative cells better than 9 hour cells. Changes in agglutinability of the cells by lectins with development suggested the possibility of receptor modification in structure or topography during development.

A direct morphological demonstration of lectin receptors on differentiated *P. pallidum* cells has been made with ferritin-conjugated pallidin (Chang *et al.,* 1977). The lectin-conjugate bound evenly over the surface of glutaraldehyde-fixed cells with no localization at the ends or sides. Thus, for both the lectin and its potential receptors, there was no evidence for regional distribution at the cell surface.

Direct binding measurements have quantified the interaction between isolated slime mold lectins and cell surface receptors on intact amoebae (Reitherman *et al.,* 1975). These studies were done by adsorbing solutions containing a fixed concentration of lectin with varying numbers of glutaraldehyde-fixed cells. The amount of bound lectin was determined by difference by measuring residual lectin in the adsorbed supernatants with a quantitative hemagglutination assay. This information allowed computation of the association constant of the lectin for each cell type as well as the total number of receptor sites per cell. Nine-hour *D. discoideum* cells bound a maximum of 4–5 x 10^5 discoidin I or discoidin II molecules per cell with an affinity of 10^9 M^{-1}. Vegetative cells, on the other hand, bound slightly less of these lectins with approximately 25-fold lower affinities. Thus in *D. discoideum*, lectins and high affinity receptors for the lectins appear on the cell surface as the cells differentiate and become cohesive. The increase in binding affinity probably contributed to the enhanced lectin-mediated agglutination with differentiation discussed above. Pallidin, like the discoidins, bound to homotypic differentiated cells with a very high affinity (4 x 10^9 M^{-1}). These very high association constants observed for the two slime molds indicated highly specific interactions of probable functional importance. In contrast, fortuitous non-biological interactions such as the binding of discoidin to erythrocytes or the binding of Con A to *D. discoideum* cells were characterized by affinity constants that were orders of magnitude lower.

The binding studies also revealed a degree of species specificity between lectins and cell surface receptors in *P. pallidum* and *D. discoideum*. Each lectin bound better to the homotypic as compared to the heterotypic cell type (8–10-fold preference). And each cell type bound homotypic lectin better than heterotypic lectin (3–20-fold advantage). It is not known to what degree binding between soluble lectins and fixed cell surfaces reflect the interactions between lectins and receptors attached to living cell surfaces. These results do, however, suggest the possibility that specific complementarity between lectins and receptors on each species may play a role in species-selective intercellular affinities.

Attempts are under way to isolate and identify the cell surface receptors defined by the agglutination and binding studies. These receptors are presumed to contain complex carbohydrates. In preliminary experiments with *P. pallidum*, an inhibitor of pallidin has been solubilized from crude membranes with detergent (Rosen *et al.,*

1976a). Some progress has been made in purifying this inhibitor, but so far it is not available in sufficient quantities for biochemical and functional analyses.

(f) *Inhibition of cohesiveness by lectin antagonists*

As reviewed above, active carbohydrate-binding proteins are present at the cell surface of aggregation-competent amoebae. In addition, cell surface receptors, to which the lectins bind with high affinity are available on the cells. A working hypothesis suggested by these results is that interactions between lectins and receptors on adjacent cells might mediate cell adhesion. If this hypothesis is correct, then appropriate concentrations of lectin inhibitors, sufficient to overcome the high affinity interaction between lectins and their receptors, should block cell adhesion.

In *P. pallidum,* this condition has been met with several inhibitors of pallidin. Specific simple saccharides such as D-galactose or lactose inhibited the EDTA-insensitive agglutination of cohesive cells much more effectively than non-specific sugars (Rosen *et al.,* 1974). However, significant effects were achieved only at high sugar concentrations (0.1 M) which were very hypertonic for the cells. To determine whether more potent inhibitors of pallidin might block cohesiveness at lower concentrations, univalent antibody against pallidin and asialofetuin were employed (Rosen *et al.,* 1976b, 1977). Microgram quantities (10^{-8} M) of either of these inhibitors did, in fact, block cohesiveness of slime mold cells, but the cells had to be exposed to hypertonic conditions (high salt or glucose) or to antimetabolites (sodium azide or 2, 4, dinitrophenol) for an inhibitory effect to be demonstrated. Inhibition was not seen under normal isotonic assay conditions.

The target of the inhibitors under the 'permissive' (hypertonic or antimetabolite) assay conditions appeared to be cell surface pallidin since:

(1) normal Fab had no effect; and
(2) chemical modification of the carbohydrate moieties of fetuin either to potentiate or inactivate it as an inhibitor of isolated pallidin had a comparable effect on its activity as an inhibitor of cohesiveness. Under the 'permissive' assay conditions, the cohesiveness of vegetative cells was always clearly less than that of differentiated cells, suggesting, but not proving, that functionally significant cell cohesion was being measured in these assays.

It is not known why successful inhibition required these extreme assay conditions. However, morphological examination of agglutinates formed under various conditions did provide some clues. Relative to the isotonic condition which did not allow inhibition, agglutinates formed under the 'permissive' conditions were much looser with much more limited areas of close cell contact. Possibly, the 'permissive' conditions eliminated or attenuated a secondary system of adhesion independent of, but complementary to a lectin-mediated system. Alternatively, the explanation might be purely quantitative; that is, the number of potential sites of cell interaction had to be restricted before the lectin antagonists at the concentrations used could have a measurable effect on agglutination.

These inhibition experiments, although highly supportive, do not establish a direct role for cell surface lectins in cell adhesion. Although temporal correlations (discussed above) support this possibility, the relationship has not been established between cell cohesiveness in artificial agglutination assays and the cellular interactions that occur in normal morphogenesis. It is possible that the assay conditions generated artifactual cellular interactions mediated by lectin. In addition, even if lectin antagonists inhibit functionally significant cellular interactions, it is not known how direct this effect is. True cell adhesion sites might neighbor lectin sites and be sterically inhibited by lectin antagonists. Or possibly, the lectins might be regulatory sites controlling the activity of true adhesion molecules. Similar possibilities were raised in the discussion of 'contact site' antigens.

(g) *Lectin system and 'contact site' antigens*

Huesgen and Gerisch (1975) have presented preliminary results on the relationship of 'contact site' antigens to the lectin system in *D. discoideum*. Partially purified CS A (160-fold) did not agglutinate formalinized sheep erythrocytes, whereas the starting detergent extract did contain hemagglutination activity. This result argued against identity of CS A with discoidin I, but not necessarily discoidin II, since this lectin does not agglutinate formalinized sheep erythrocytes. Since CS A contained associated carbohydrate, it is possible that the antigens are identical to the lectin receptors or possibly a lectin–receptor complex. The purification and characterization of CS A and of the discoidin receptors, which is in progress, should resolve this question.

7.4 CONCLUSION

Because of their favorable experimental properties, the cellular slime molds have been employed for an extensive series of investigations of intercellular adhesion. It has been established that the cell surface of cellular slime molds undergoes specific changes as the cells differentiate from a vegetative to a cohesive form, and that some of these changes are probably important in cell cohesion. There is considerable evidence that certain cell surface antigens collectively referred to as 'contact sites' A play some role in cell cohesion; but whether or not these molecules are direct cohesion molecules remains to be determined. It is also clear that developmentally regulated carbohydrate-binding proteins (lectins) and complementary oligosaccharide receptors are present on the surface of cohesive cells and that specific lectin antagonists block cell adhesion under certain conditions. Whereas the role of this system in cell cohesion has not been formally proved, as reviewed above, it is an attractive candidate for a cohesion system not only because of the experimental evidence but also because of the plausibility of the proposed mechanism.

In addition to further direct studies with slime molds, related studies with other systems may provide additional information about the role of cell surface lectins in cell cohesion. A discussion of the application of the possible importance of lectins in cell adhesion in other systems has been presented (Barondes and Rosen, 1976). Recent evidence for developmentally regulated lectins in embryonic chick muscle (Nowak *et al.*, 1976) that are present on the cell surface (Nowak *et al.*, 1977) encourages the speculation that cell surface lectin involvement in cell cohesion may prove to be a general biological phenomenon.

REFERENCES

Alcantara, F. and Bazill (1976), *J. gen. Microbiol.*, **92**, 351−368.

Aldrich, H.C. and Gregg, J.H. (1973), *Exp. Cell Res.*, **81**, 407−412.

Alexander, S., Brackenbury, R. and Sussman, M. (1975), *Nature*, **254**, 698−699.

Ashworth, J.M. and Sussman, M. (1967), *J. biol. Chem.*, **242**, 1696−1700.

Barondes, S.H. and Rosen, S.D. (1976), In: *Neuronal Recognition*, (ed. Barondes, S.H.), Plenum Press, New York, pp. 331−356.

Beug, H., Gerisch, G., Kempff, S., Riedel, V. and Cremer, G. (1970), *Exp. Cell Res.*, **63**, 147−158.

Beug, H., Gerisch, G. and Muller, E. (1971), *Science*, **173**, 742−743.

Beug, H., Katz, F.E. and Gerisch, G. (1973a), *J. Cell Biol.*, **56**, 647−658.

Beug, H., Katz, F.E., Stein, A. and Gerisch, G. (1973b), *Proc. natn. Acad. Sci. U.S.A.*, **70**, 3150−3154.

Bonner, J.T. (1947), *J. exp. Zool.*, **106**, 1−26.

Bonner, J.T. (1949), *J. exp. Zool.*, **110**, 259−271.

Bonner, J.T. (1950), *Biol. Bull.*, **99**, 143−151.

Bonner, J.T. (1957), *Q. Rev. Biol.*, **32**, 232−246.

Bonner, J.T. (1967), *The Cellular Slime Molds*, 2nd edn., Princeton University Press, New Jersey.

Bonner, J.T. (1970), *Proc. natn. Acad. Sci. U.S.A.*, **65**, 110−113.

Bonner, J.T. (1971), *A. Rev. Microbiol.*, **25**, 75−92.

Bonner, J.T. and Adams, M.S. (1958), *J. Embryol. exp. Morphol.*, **6**, 346−356.

Bonner, J.T., Barkley, D.S., Hall, E.M., Konijn, T.M., Mason, J.W., O'Keefe, G. and Wolfe, P.B. (1969), *Dev. Biol.*, **20**, 72−87.

Bonner, J.T., Hall, E.M., Noller, S., Oleson, F.B. and Roberts, A.B. (1972), *Dev. Biol.*, **29**, 402−409.

Bonner, J.T., Sieja, T.W. and Hall, E.M. (1971), *J. Embryol. exp. Morph.*, **25**, 437−465.

Brenner, M. (1977), *J. biol. Chem.*, **252**, 4073−4077.

Chang, C.-M., Reitherman, R.W., Rosen, S.D. and Barondes, S.H. (1975), *Exp. Cell Res.*, **95**, 136−142.

Chang, C.-M., Rosen, S.D. and Barondes, S.H. (1977), *Exp. Cell Res.*, **104**, 101−109.

Cohen, M.H. and Robertson, A. (1971), *J. theor. Biol.*, **31**, 101−118.

Coston, M.B. and Loomis, W.F. (1969), *J. Bact.*, **100**, 1208−1217.

Curtis, A.S.G. (1967), *The Cell Surface: Its Molecular Role in Morphogenesis*, Academic Press, London.

Darmon, M., Brachet, P. and Pereira Da Silva, L.H. (1975), *Proc. natn. Acad. Sci. U.S.A.*, **72**, 3163–3166.

Darmon, M. and Klein, C. (1976), *Biochem. J.*, **154**, 743–750.

Ellingson, J.S. (1974), *Biochim. biophys. Acta*, **337**, 60–67.

Farnsworth, P.A. and Wolpert, L. (1971), *Nature*, **231**, 329–330.

Firtel, R.A. and Bonner, J.T. (1972), *Dev. Biol.*, **29**, 85–103.

Frazier, W.A., Rosen, S.D., Reitherman, R.W. and Barondes, S.H. (1975a), *J. biol. Chem.*, **250**, 7714 7721.

Frazier, W.A., Rosen, S.D., Reitherman, R.W. and Barondes, S.H. (1975b), In: *Cell Surface Receptors*, pp. 57–66, (Bradshaw, R.A., Frazier, W.A., Merrell, R.C., Gottlieb, D.I. and Hogue-Angeletti, R.A. eds,), Plenum Press, New York.

Garrod, D.R. (1974), *Arch. Biol.*, **85**, 7–31.

Garrod, D.R. and Forman, D. (1977), *Nature*, **265**, 144–146.

Garrod, D.R. and Gingell, D. (1970), *J. Cell Sci.*, **6**, 277–284.

Geltowsky, J.E., Siu, C.-H. and Lerner, R.A. (1976), *Cell*, **8**, 391–396.

Gerisch, G. (1961), *Exp. Cell Res.*, **25**, 535–554.

Gerisch, G. (1965), *Publ. Wiss. Filmen*, **1A**, 265–278.

Gerisch, G. (1968), *Curr. Top. Dev. Biol.*, **3**, 157–197.

Gerisch, G. (1975), In: *Surface Membrane Receptors*, pp. 67–72 (Bradshaw, R.A., Frazier, W.A., Merrell, R.C., Gottlieb, D.I. and Hogue-Angeletti, R.A. eds,), Plenum Press, New York.

Gerisch, G. (1976), *Cell Differ.* **5**, 21–25.

Gerisch, G., Beug, H., Malchow, D., Schwarz, H. and Stein, A.V. (1974), *Miami Winter Symp.* **7**, 49–66.

Gerisch, G., Fromm, H., Huesgen, A. and Wick, U. (1975), *Nature*, **255**, 547–549.

Gerisch, G. and Huesgen, A. (1976), *J. Embryol. exp. Morph.*, **36**, 431–442.

Gerisch, G. and Malchow, D. (1976), *Adv. Cycl. Nuckot. Res.*, **7**, 49–68.

Gerisch, G., Malchow, D., Riedel, V., Muller, E. and Every, M. (1972), *Nature New Biol.*, **235**, 90–92.

Gillette, M.U., Dengler, R.E. and Filosa, M.F. (1974), *J. exp. Zool.*, **190**, 243–248.

Gillette, M.U. and Filosa, M.F. (1973), *Biochem. biophys. Res. Comm.*, **53**, 1159–1166.

Goidl, E.A., Chassy, B.M., Love, L.L., and Krichevsky, M.I. (1972), *Proc. natn. Acad. Sci. U.S.A.*, **69**, 1128–1130.

Grabel, L. and Farnsworth, P.A. (1977), *Exp. Cell Res.*, **105**, 285–289.

Grabel, L. and Loomis, W.F. (1977), In: *Development and Differentiation in Cellular Slime Molds*, (ed., Cappucinelli, P.,) Elsevier, Amsterdam.

Gregg, J.H. (1956), *J. gen. Physiol.*, **39**, 813–820.

Gregg, J.H. (1971), *Dev. Biol.*, **26**, 478–485.

Gregg, J.H. and Nesom, M.G. (1973), *Proc. natn. Acad. Sci. U.S.A.*, **70**, 1630–1633.

Gregg, J.H. and Trygstad, C.W. (1958), *Exp. Cell Res.*, **15**, 358–369.

Hoffman, S. and McMahon, D. (1977), *Biochim. biophys. Acta*, **465**, 242–249.

Huesgen, A. and Gerisch, G. (1975), *FEBS Letters*, **56**, 46–49.

Jacobson, A. and Lodish, H. (1975), *A. Rev. Genet.*, **9**, 145–183.

Jones, M. and Robertson, A. (1976), *J. Cell Sci.*, **22**, 41–47.

Kawai, S. and Takeuchi, I. (1976), *Dev. Growth Differ.*, **18**, 311–317.

Killick, K.A. and Wright, B.E. (1974), *A. Rev. Microb.*, **28**, 139–166.

Klein, C. and Darmon, M. (1976), *Proc. natn. Acad. Sci. U.S.A.*, **73**, 1250–1254.

Konijn, T.M., van de Meene, J.G.C., Bonner, J.T. and Barkley, D.S. (1967), *Proc. natn. Acad. Sci. U.S.A.*, **58**, 1152–1154.

Leach, C.K., Ashworth, J.M. and Garrod, D.R. (1973), *J. Embryol. exp. Morph.*, **29**, 647–661.

Lee, K.-C. (1972a), *J. Cell Sci.*, **10**, 229–248.

Lee, K.-C. (1972b), *J. Cell Sci.*, **10**, 249–265.

Liener, I.G. (1976), *A. Rev. Plant. Physiol.*, **27**, 291–319.

Long, B.H. and Coe, E.L. (1974), *J. biol. Chem.*, **249**, 521–529.

Loomis, W.F. (1969), *J. Bact.*, **100**, 417–422.

Loomis, W.F. (1975), *Dictyostelium discoideum, A Developmental System,* Academic Press, New York.

Maeda, Y. and Takeuchi, I. (1969), *Dev. Growth Differ.*, **11**, 232–245.

Malchow, D. and Gerisch, G. (1973), *Biochem. biophys. Res. Comm.*, **55**, 200–204.

Malchow, D. and Gerisch, G. (1974), *Proc. natn. Acad. Sci. U.S.A.*, **71**, 2423–2427.

Malchow, D., Nagele, B., Schwarz, H. and Gerisch, G. (1972), *Eur. J. Biochem.*, **28**, 136–142.

Mercer, E.H. and Shaffer, B.M. (1960), *J. biophysic. biochem. Cytol.*, **7**, 253–263.

Molday, R., Jaffe, R. and McMahon, D. (1976), *J. Cell Biol.*, **71**, 314–322.

Nanjundiah, V. and Malchow, D. (1976), *J. Cell Sci.*, **22**, 49–58.

Newell, P.C. (1971), *Essays in Biochem.*, **7**, 87–126.

Newell, P.C., Longlands, M. and Sussman, M. (1971), *J. Mol. Biol.*, **58**, 541–554.

Newell, P.C., Franke, J. and Sussman, M. (1972), *J. mol. Biol.*, **63**, 373–382.

Nicolson, G.L. (1974), *Int. Rev. Cytol.*, **39**, 89–190.

Nowak, T.P., Haywood, P.L. and Barondes, S.H. (1976), *Biochem. biophys. Res. Comm.*, **68**, 650–657.

Nowak, T.P., Kobiler, D., Roel, L.E. and Barondes, S.H. (1977), *J. biol. Chem.* (In press).

Olive, L.S. (1974), *The Mycetozoans,* Academic Press, New York.

Pan, P., Bonner, J.T., Wedner, H.J. and Parker, C.W. (1974), *Proc. natn. Acad. Sci. U.S.A.*, **71**, 1623–1625.

Raper, K.B. (1940), *J. Elisha Mitchell Sci. Soc.*, **56**, 241–282.

Raper, K.B. (1941), Third Growth Symposium, *Growth*, **5**, 41–76.

Raper, K.B. (1973), In: *The Fungi*, Vol. IVB, (Ainsworth, G.C., Sparrow, F.K. and Sussman, A.S. eds,), pp. 9–36, Academic Press, New York.

Raper, K.B. and Thom, C. (1941), *Am. J. Bot.*, **28**, 69–78.

Reitherman, R.W., Rosen, S.D., Frazier, W.A. and Barondes, S.H. (1975), *Proc. natn. Acad. Sci. U.S.A.*, **72**, 3541–3545.

Robertson, A. and Cohen, M.H. (1972), *A. Rev. Biophys. Bioeng.*, **1**, 409–464.

Robertson, A., Drage, D.J. and Cohen, M.H. (1972), *Science*, **175**, 333–335.

Robertson, A. and Grutsch, J. (1974), *Life Sci.*, **15**, 1031–1043.

Rosen, S.D. (1972), A Possible Assay for Intercellular Adhesion Molecules, Ph. D. Thesis, Cornell University.

Rosen, S.D., Chang, C.-M. and Barondes, S.H. (1977), *Dev. Biol.*, (In press).

Rosen, S.D., Haywood, P.L. and Barondes, S.H. (1976a), *J. Cell Biol.*, **70**, 133a.

Rosen, S.D., Haywood, P.L. and Barondes, S.H. (1976b), *Nature,* **263,** 425–427.
Rosen, S.D., Kafka, J.A., Simpson, D.L. and Barondes, S.H. (1973), *Proc. natn. Acad. Sci. U.S.A.,* **70,** 2554–2557.
Rosen, S.D., Reitherman, R.W. and Barondes, S.H. (1975), *Exp. Cell Res.,* **95,** 159–166.
Rosen, S.D., Simpson, D.L., Rose, J.E. and Barondes, S.H. (1974), *Nature,* **252,** 128 and 149–151.
Rubin, J. and Robertson, A. (1975), *J. Embryol. exp Morph.,* **33,** 227–241.
Runyon, E.H. (1942), *Collecting Net,* **17,** 18.
Shaffer, B.M. (1953), *Nature,* **171,** 975.
Shaffer, B.M. (1957a), *Am. Nat.,* **91,** 19–35.
Shaffer, B.M. (1957b), *Q. J. Micros. Sci.,* **98,** 393–405.
Shaffer, B.M. (1961), *J. exp. Biol.,* **38,** 833–849.
Shaffer, B.M. (1964), In: *Primitive Motile Systems in Cell Biology,* (Allen R.D. and Kamiya, N. eds,), pp. 387–405, Academic Press, New York.
Shaffer, B.M. (1972), *Nature,* **255,** 549–552.
Sharon, N. and Lis, H. (1972), *Science,* **177,** 949–959.
Simpson, D.L., Rosen, S.D. and Barondes, S.H. (1974), *Biochemistry,* **13,** 3487–3493.
Simpson, D.L., Rosen, S.D. and Barondes, S.H. (1975), *Biochim. biophys. Acta,* **412,** 109–119.
Siu, C.-H., Lerner, R.A., Firtel, R.A. and Loomis, W.F. (1975), In: *Pattern Formation and Gene Regulation in Development,* (McMahon, D. and Fox, C.F. eds,), pp. 129–134, W.A. Benjamin, Palo Alto.
Siu, C.-H., Lerner, R.A., Ma, G., Firtell, R.A. and Loomis, W.F. (1976), *J. mol. Biol.,* **100,** 157–178.
Smart, J.E. and Hynes, R.O. (1974), *Nature,* **251,** 320–321.
Smart, J.E. and Tuchman, J.E. (1976), *Dev. Biol.,* **51,** 63–76.
Sonneborn, D.R., Sussman, M. and Levine, L. (1964), *J. Bact.,* **87,** 1321–1329.
Steinberg, M.S. (1964), In: *Cellular Membranes in Development,* (Locke, M., ed.), pp. 321–366, Academic Press, New York.
Sussman, M., and Boschwitz, Ch. (1975), *Dev. Biol.,* **44,** 362–368.
Sussman, M. and Brackenbury, R. (1976), *A. Rev. Plant Physiol.* **27,** 229–265.
Takeuchi, I. (1963), *Dev. Biol.,* **8,** 1–26.
Takeuchi, I. (1969), In: *Nucleic Acid Metabolism, Cell Differentiation and Cancer Growth,* (Cowdry, E.V. and Seno, S., eds), pp. 297–304, Pergamon Press, New York.
Takeuchi, I. and Yabuno, K. (1970), *Exp. Cell Res.,* **61,** 183–190.
Town, C.D., Gross, J.D. and Kay, R.R. (1976), *Nature,* **262,** 717–719.
Tuchman, J.E., Smart, J.E. and Lodish, H.F. (1976), *Dev. Biol.,* **51,** 77–85.
Von Dreele, P.H. and Williams, K.L. (1977), *Biochim. biophys. Acta,* **464,** 378–388.
Weeks, G. (1973), *Exp. Cell Res.,* **76,** 467–470.
Weeks, G. (1975), *J. biol. Chem.,* **250,** 6706–6710.
Weeks, C. and Weeks, G. (1975), *Exp. Cell Res.,* **92,** 372–382.
West, C. and McMahon, D. (1977), *J. Cell Biol.,* **74,** 264–273.

Wurster, B., Pan, P., Tyan, G.-G. and Bonner, J.T. (1976), *Proc. natn. Acad. Sci. U.S.A.*, **73**, 795–799.

Yabuno, K. (1970), *Dev. Growth Differ.*, **12**, 229–239.

Yabuno, K. (1971), *Dev. Growth Differ.*, **13**, 181–190.

Index